Towards a Newer World

Towards a Newer World

B. R. SEN

Published by

TYCOOLY INTERNATIONAL PUBLISHING LTD.

DUBLIN

Published by:
Tycooly International Publishing Ltd.,
6 Crofton Terrace,
Dun Laoghaire,
Co. Dublin, Ireland
Telephone: (+353-1) 800245/6
Telex: 30547 SHCN EI

First edition 1982

Typeset by Healyset Ltd., Dublin.
Printed by Irish Elsevier Printers Ltd., Shannon, Co. Clare, Ireland.
Cover by Printset and Design, Dublin.

ISBN 0 907 567 26 6 Hardback
ISBN 0 907 567 27 4 Softcover

Contents

Part I: The Changing Times

Part II: Freedom from Hunger

Part I

The Changing Times

To my wife Chiroprova
remembering
her patience, understanding and tolerance
all through the years
and her
all-encompassing compassion

Preface

IT IS NOT my intention to attempt an autobiography. I am an average person and the life of such a person is the life of any other person. In one respect I could say that there is a difference. I have had unique opportunities of seeing the world from many different angles and from very privileged vantage points in my role as civil servant, diplomat and United Nations official.

My life spans the present century. With the two World Wars and the Bolshevik Revolution challenging the whole existing social order and promising a new world, forces of immeasurable possibility and enormous danger have been released. The scientific discoveries since the Second World War, particularly nuclear weapons, have thrown a dark shadow over man's very existence.

How does a man, moulded in the age-old traditions of India with its proud past, react and respond to changes of such vast dimensions? This, rather than a conventional autobiography, is the story that I will attempt. In doing so, I have kept in mind what R.A. (Lord) Butler so rightly said in his autobiography: "Some degree of egoism, some belief in one's star is essential to autobiography. An author may overdo this. He may not only be too ready to praise or exaggerate his own talents and successes, but too prone to drop a veil over his humiliations and eccentricities ..." I will keep the latter part of his remarks in view, as a warning, as I proceed.

Chapter 1

My Early Days

Growing-up

THIS YEAR I exceed the Biblical span of life by a decade. I was born at the turn of the century and was fifth in a family of nine brothers and sisters. We were a closely knit family, playing, laughing, quarrelling and falling ill under the watchful eye of a mother who enveloped us with a warm love. My father was a more distant figure. He was a doctor, always busy and away from home. We surrounded him with an aura of awe, a kind of respectful fear.

Some evenings at home I remember in a special way. I used to lie on my back on a mat spread out in the courtyard under the full moon listening to my mother telling stories from the *Ramayana*. I was always moved by the story of the forcible abduction of Sita by Ravana and how Sita was traced by Rama with help from Garuda. In quiet moments I can still fantasize the fearful Ravana in the far away Lanka!

The first day I was taken to kindergarten is still a vivid memory. My teacher picked me up and guiding my outstretched hand pointed out north, south, east and west, and asked me to repeat the exercise. Later, as I began to go to school by myself, I was given a rather large man's umbrella as it always seemed to rain. As I was no more than four years old, my mother used to say that whenever she saw an umbrella coming towards our house she knew I had to be somewhere inside it! On the way to school there was a stile and a wire-fence to cross which was something of a problem for me with my load of books and the umbrella. Once, as I was climbing the stile, there was a flash of lightning and a roll of thunder. I was really frightened. I sobbed and cried till help came.

I was very fond of my father's elder sister, an old lady who lived in an attached cottage. She used to wear the saffron clothes of a widow. Every morning she sat down cross-legged on a mat to do *puja*. She had a *thali* (plate) in front of her with some flowers (which I was often delegated to collect from the garden), a little unhusked rice, some shoob grass, sandal wood paste and an earthen lamp. I was her constant companion and sat in front of her, fascinated, as she busied herself with her various chores. One morning I went into her room to find it empty. She was gone, never to return. It was explained to me that she had gone to heaven. I remember the shock and the feeling of emptiness. I couldn't understand how a kindly person like her could suddenly disappear and never come back again.

My entire childhood was overshadowed by the prolonged illness of my eldest brother. The only cure for him, we were told, was a change of air. For two consecutive summers the whole family, except my father who had to stay back to look after his practice, travelled to the dry climate of Chhotanagpur in Behar. I was no more than four years old. We travelled by river (the Brahmaputra) all the way from Dibrugarh in the north-eastern part of Assam to Calcutta, and then made the rest of the journey by rail. The boat was a slow, flat-bottomed paddle-steamer which made many stops on the way. As deck passengers we were free to roam the ship. We ran down to the engine room to see the powerful engines at work and back up again to look over the side and watch the water being thrashed and churned by the huge paddles. We were favourites with the crew who enjoyed our frolics because they missed their own children on these long journeys. At every stop, little boats carrying passengers or with provisions for sale used to crowd around our boat. It was wildly exciting for a child, with people shouting, gesturing and crying. At night, the huge searchlights of the steamer lit the way ahead in the tortuous river, and a vast array of flies were picked up in the powerful beams. The hustle and bustle of those journeys was etched forever in my mind. Perhaps these journeys planted the seed of restlessness that was to develop in later life into an urge to travel around the world.

Our summer home in those days was near a police station. One morning I heard a commotion at the station and I saw a father holding his dead son in his arms. He was crying his heart out. The boy had just died of snake bite, and looked blue and still. This was the first time I saw what they called death at close quarters.

I was soon to experience my own brother's death. He was very ill. There was silence in the house and I had gone to bed. I woke up to the sound of voices and cries. My dead brother's body was lying in the middle of the hall surrounded by many students from the Medical School where my father taught. My mother's grief was uncontrollable. My father sat near the head of the bed. For the first time in my life I saw him crying and calling to God for consolation. We were sent back to bed but

I couldn't sleep. What I had seen and heard was traumatic. Night wore on until I could finally see the first light of dawn peeping through under the thatch on the roof. What ease and comfort the light of dawn brought to me!

My brother's death was a great blow to the whole family. We loved him and envied his growing into manhood. My mother continued to be very upset and cried often. Many times as I returned from the kindergarten I could hear the sound of her sobs from a distance. It was her way to find relief. As we grew older she mentioned him less and less in our presence, but her grief was deep and lasting. We noticed that she never again touched the utensils that he used at mealtimes. His death was a sad ending to my childhood years.

My boyhood was comparatively happier but it had its own trials and tribulations. I was sent to a joyless school where I encountered many bullies. I found the bigger boys so oppressive that I used to wait for the teacher to come before I would go into the classroom. We had many Gurkha boys in the school from the Gurkha regiments stationed in the town. I soon became great friends with a Gurkha lad who assumed the role of my protector. Once, when someone tried to molest me he took up a long pencil and hit him on the head drawing blood. I had fewer problems after that.

The teachers were unimpressive as educationalists. I remember one who was always drunk in the middle of the day. All he did by way of class work was to have two of his chosen boys attend to his comforts by massaging his head and hands! Another had a passion for caning, often with no justification. A third had a stiff and painful leg and went into an uncontrollable rage at the slightest provocation. Another had an ambition to be an orator in English. On annual prize-giving days, he would stand up boldly behind his chair to make a speech and off he went, round and round the chair shaking like an angry tiger in a cage, searching for momentous words that never came. We learned little enough from all our teachers. My constant thought was how to avoid school. On rainy days, I was one of those who stood under a gushing rain pipe to get thoroughly wet. Then we were allowed to leave and change our sodden clothes. I seldom returned. Outside of school hours, I was happy with a few friends and my elder brother. Once he and I were playing at the Russo-Japanese war, then the main topic among the elders, throwing things at each other from either side of a long ditch. I was the one to get the first hit in, opening a gash in his head. I pleaded with him not to tell at home. When he presented himself to mother covered in blood, I knew I was in trouble but, with real nobility, he said he had cut himself on a wire fence.

As I grew older, I took to sports like hockey, football and cricket. Though I was quite a small boy, I often won a place on the school team and played in matches against some of the regimental teams. I was often the carrier of the 'letter of challenge' to the regimental officers.

When I was twelve the whole family was sent to Dacca, our original home district. I remember the feeling of desolation on being so suddenly uprooted from my familiar surroundings and sent to Dacca. I was admitted into the prestigious Dacca Collegiate School. The teachers were much better than in my first High School, and the school discipline was good. Yet, I found it hard to adjust. I couldn't make friends and the city was so enormous. I was thoroughly unhappy for the six months I spent there. One interesting event I remember from that time was the parading of what were called "the Dacca Conspiracy Case Prisoners" through the streets of the city. They were driven each day in an open carriage, more than a dozen in number, from the prison to the City Court House for their trial. The carriage was drawn by two magnificent black horses which was a sight in itself. Was this spectacular parade intended to demonstrate the invincible power of the British Raj? If so, how mistaken the Raj was! The little boy of twelve and the thousands who lined the streets to view this procession were not impressed by the invincibility of the Raj but paid a silent homage to those brave men who had defied Lord Curzon and all he stood for.

I returned from Dacca to face three more years in my old familiar school. I developed an interest in study. I soon came to the top of the class and became an avid reader. I loved the Bengali novels of Bankim Chatterjee and often read the poetry of Tagore. I got to know some of the writings of Romesh Dutt. These first two authors have remained my constant companions ever since. Years later I was very glad of the opportunity to help bring out an authoritative edition of all the works of Bankim Chatterjee. Among the English authors, Scott's novels were as much as I could manage.

I matriculated at fifteen, somewhat battered, but with my zest for learning undiminished and ready to go to college. I was to leave home for an entirely new adventure in the biggest city in India — Calcutta. I was to be admitted into the Scottish Churches College, one of the premier colleges in the country, administered by Scottish missionaries. I left my boyhood years behind and looked forward to the prospect of a new kind of life and new studies.

Even as a boy I wanted to be in the Indian Civil Service (ICS). In the district where we lived, the ICS was represented by District Officers or assistants to the District Officers. They were the final appeal in all matters of everyday life. I sometimes saw them on cricket or polo fields. One was a Bentinck, probably a descendant of Lord Bentinck, Governor-General of India, who had left behind a great name as a humanitarian and educationalist. I used to see him posturing round the stumps before every ball, fearful almost that he might be bowled out. Another District Officer was called Evans. I remember him specially, because while I was walking across the school playground I saw a small white ball suddenly bouncing towards me. I looked around in surprise and saw the Officer flourishing a queer-looking stick coming towards me. This was my first introduction to golf! And it was the nearest I came for many years to the Indian Civil Service.

My respect for the power and authority enjoyed by the people in the Service grew with the years and I was determined to make the ICS my future career. Before I had left school my father had told me that I could have a scholarship to the Medical College but I had no interest in the medical profession, though I knew my father enjoyed great respect and prestige as a doctor. I do not know if he was disappointed when I told him that the medical profession was not for me.

When I went to college the ICS was always in the forefront of my mind though I knew that the competitive examination which was held in London was one of the hardest in the world. Top graduates of Oxford and Cambridge, and other British and Irish universities usually competed for that examination. It was easier for them since it was in line with the courses they followed at their universities. I was also aware that, so far, I had not distinguished myself sufficiently in my academic career. Yet, my one-track mind would not admit defeat.

However, I must not run ahead of time. I entered college in May 1914. The Great War broke out and dragged on. Only when the newspapers, like *The Statesman,* started carrying page after page of casualty-lists every day did the enormity of the losses in human life become a reality for us. Several of our own people had gone to war with the Indian contingents. The surrender of General Townsend in Mesopotamia to the Turks was followed by an outbreak of typhus in which many Indians died. The war overshadowed every aspect of our lives.

In college we had a fine Scotsman, Dr. Cameron, who taught English Literature. His method of teaching was to read English poetry without explanation but with great feeling. It was a novel experience, how reading in itself, with the proper emphasis and emotion, could reach into the minds and hearts of the listeners. He read Keats' Odes, which remain to me the quintessence of poetry. I listened to him, entranced. He seemed to kindle new light in my innermost being. The lines in the *Ode to a Nightingale* left me breathless with their beauty. These readings, in a way, transformed my personality. I felt different, more grown up, ushered into a new world of romance.

Another lecturer who left a permanent impression on me was Dr. Haridas Bhartacharya who was a brilliant scholar. Philosophy was his subject. He had a phenomenal memory. As soon as he came into the class, he would begin his lecture without any notes or books before him. This would go on for an hour. Next day he would enquire where he had stopped the previous day and immediately resume his lecture without any break in thought. He had no ambitions in life except to teach philosophy.

One or two other events come to mind from my college days. Sir Ashutosh Mookherjee, a distinguished Judge of the High Court and, more importantly, Vice-Chancellor of the Calcutta University, visited our College. He was an outstanding figure in public life. He came into prominence because of his defiance of the Chancellor of the University, Lord Lytton, then Governor of Bengal, over some University matter. The day-to-day running of the University was in his hands. He was greatly feared and respected by everyone who had contact with him. When he came to our college, scores of boys rushed up to him and struggled to take the dust off his feet — the Indian way of showing respect — but this unseemly struggle left me cold and I had nothing to do with it. I held myself at a distance.

Another distinguished visitor was Professor Jayaswal, a scholar of ancient Indian history, a noted anthropologist and Curator of the Patna Museum. It was his researches which revealed that the earliest incidence of election by secret voting took place in the Buddhist monasteries in north-west India at least 2,000 years ago. The monks used *shalakas* (coloured needles) when they cast their secret votes in the election of office bearers.

A large gathering assembled to hear him. He spoke of the coming of the Aryans to India and their dispersal and mixture with the indigenous populations. He specifically referred to north-eastern India, including Bengal. When someone asked if he could find anyone in the assembly who still retained the characteristics of an Aryan in appearance, he looked around and pointed to me! That was the last thing I expected. I was greatly confused and embarrassed.

But those were happy years. I was entirely free to follow my own pursuits, ignoring the social, political and various activities which I found distracting. I did not even join in the college sports, though when I went home I resumed tennis, which was my favourite sport. I used to take long walks — as I have done ever since — and I was happy to be alone. It was rather a lonely life, but it suited my temperament. I graduated in 1918 with a B.A. degree in English Language and Literature with honours. I was ready to go to England and sit for the ICS examination. But the war was still on and I had to be patient and wait. I was very frugal in my habits, because I realised that my father was not wealthy and I also thought of my brothers and sisters who were growing up and who also needed his support.

I joined the University Post-Graduate College for my M.A. degree in English Literature and Language with Old English as my special subject. Professor Suniti Kumar Chatterjee, one of the most brilliant philologists India has produced, was my lecturer in Anglo-Saxon. His command of the languages of European countries was something of a marvel. He was extraordinarily brilliant. When I was Ambassador to Mexico and the USA, I inducted Professor Chatterjee as my Honorary Cultural Attaché to give him an opportunity to visit Mexico and study the old Maya and Inca civilizations. He was an indefatigable traveller and his love of knowledge was very deep.

The professor of English poetry was Dr. Mahim Ghosh, a noted poet

the sun broke through I began to improve and the whole world changed. I stayed on board nursing myself while other passengers went ashore to see the magnificent sights of Cairo.

The majority of the passengers were Indians and, for the first time, I met Indians from different parts of the country; I realized how provincial my attitudes and outlook had been up to this. Being with them gave me a wonderful sense of community with the whole Indian nation. We disembarked at Marseilles to be told that the sea in the Bay of Biscay was rough with high waves. The thought of another spell of sea-sickness scared me and I very willingly took the train via Paris to Boulogne. Some of my friends got off at Paris to taste, as they said, the pleasures of that city, but my only thought was to reach my destination.

We arrived at Charing Cross in the heart of London around midnight and were met by the Social Secretary of the official Indian Hostel, who took us to the hostel at Cromwell Road. Even though we were wearing winter clothes, the cold on that first October night in England seemed to penetrate one's bones. We sat huddled around an open fire all night. But the journey was over and Oxford, full of history and noble tradition, was beckoning to me.

Salad Days at Oxford

I SPENT a week or ten days in London briefing myself on Oxford and I also bought some warm clothes for the English winter. As I prepared to leave for Oxford, the General Strike (the first of its kind in England) took place, bringing train services to a halt. I do not now quite recollect how I arrived there but probably it was by a bus improvised for the occasion. I do recollect, however, my first sight of Oxford: the stone towers glinting in the sun.

I arranged lodgings in the Mitre Hotel — I do not know if it exists now — and began to enquire about admission to an Oxford College. I had been advised by the Indian High Commission authorities in London to make contact with St. Catherine's College. It did not have boarding facilities of its own for students, like the other colleges, and foreign students coming without notice had a better chance of being enrolled but without any offer of accommodation. I went with my papers to see Dr. Bernard Baker, the Head of St. Catherine's. He looked at me, asked a few general questions and, after I said that I had come all the way from India, he accepted me forthwith. I now went to find my lodgings. I remember I took a hansom — a two-wheeled horse-drawn cab with the driver sitting at the rear, the reins going over the roof. I hadn't seen a carriage like this before. It was raining, a usual feature of Oxford as I discovered later, and I got thoroughly wet getting in and out of the cab. I was turned away by several landladies — Oxford landladies were still not quite used to Asians in those days — but at last I found what I wanted at No. 7 Warwick Street, down along the Iffley Bridge. The house was run by a widow who had three daughters — two of them teenagers, the younger being twelve. They received me warmly. During my three-year stay in Oxford, this was to be my home. Like all houses in those days, it was unheated except for a fire for some hours in the evenings in the drawing room. I remember the ghastly sensation of getting into the cold bed at night; it was like plunging into an icy lake. I plucked up my courage each winter night and dived in! The cold affected me all during my stay at Oxford. One day, while cycling to college, my hands got benumbed and I remember falling off the cycle on the Iffley Bridge, scattering my books all over the place.

It took me the whole Michaelmas Term (October to December) to get adjusted to the new environment. I took History and Economics as my main subjects. My tutor used to ask us to prepare weekly essays, having first listed the books to be read. Attendance at classes was entirely at the discretion of the students; there was no roll-call or periodic examinations to check the progress of students. The tutorial system with its weekly essays was the sheet-anchor of the Oxford system.

Some of the old Indian resident-students in the University were very kind to me. Among them I remember Shastri (Balliol), Suntharalingam (Balliol), Mookherjee (Balliol) and Menon (St. Johns). It was Shastri who first introduced me to the ritual of dining at college. Suntharalingam, from Ceylon, got a First in Mathematics. Mookherjee, who came from my home city, Calcutta, was lecturer in Classics in Balliol all through the war and was a man of unusual brilliance. Within a short time I began to feel more at home in my surroundings. Welingkar, my next-door-neighbour in Warwick Street, became my closest friend. He read for a degree in English Literature, though he was primarily a Sanskrit scholar, and on his return to India pursued research in the Sanskrit language as his main interest.

Among the other students who joined Oxford at the same time were Dr. Banerjee-Shastri, a distinguished Sanskrit scholar; Abdur Rahman Siddiqui, a firebrand who during the war had taken an active part in the Khilafat movement aimed at preserving the Turkish empire and maintaining the Sultan of Turkey as the *Khalif* (the Head) of the Islamic world; Tulsi Goswami from Calcutta, an aristocrat whose aim was to model his political career on the British aristocrats, making Oxford a stepping stone to future greatness — he was a very generous man, wholesome to his fingertips; M.C. Chagla, a keen follower of Mohammad Ali Jinnah who wanted a special position for the Moslems in the Montagu-Chelmsford Reforms, which the Indian leaders had assembled in London to discuss; K.P.S. Menon, a brilliant student who was reputed never to have come second in any examination and who later topped every examination at Oxford, including, I believe, the All Souls Fellowship which he did not accept, having at that time taken his place in the Indian Civil Service. All these contemporaries of mine came to distinguish themselves in later life.

We had quite strong links with our fellow colleagues in Cambridge. Among Cambridge students I especially remember Dilip Roy, son of a noted poet and playwright from Bengal. He was a fine singer whose contribution at musical soirées was a feature of our social life. Another Cambridge student was Subhas Bose, who had joined the ICS the previous year but resigned soon after, reportedly taking offence at a remark in an old Manual of Instructions to new entrants which said that *sayces* (grooms) were not to be trusted to feed the horses properly (implying that they would keep some of the money for themselves). He was truly a great patriot built in the mould of a giant as his later life was to show to the world. I had known him when we were fellow students in Calcutta. He was expelled from the Presidency college after assaulting Professor Oaten who made some offensive remarks about Indians in his class. On his expulsion he asked for, and got, admission into the Scottish Churches College where I was also studying. He was a brilliant student, apart from being a firebrand when national pride was at stake.

Among the early visitors to Oxford from India, I remember Sarojini Naidu whose poetry in English had enchanted me while still in College in India. She spoke to the Indian students one evening choosing "Ideals and Illusions" as her subject, which gave full scope to her poetic genius. I listened to her spellbound. When it was all over, I tried to recollect what she had said and list it point by point. But the spell was broken and I could not remember it that way because I was listening to a poetess who was weaving a web of atmosphere and magic.

Acharya Jagadish Chandra Bose, the famous physicist and botanist from Calcutta University, whose theories about the behaviour of plants had caught the imagination of our generation, was also among the early visitors to Oxford. We went to hear him lecture. There was an instrument on the raised platform to show how the sap rose in a plant under stimulation. When a scientist from our University went to examine the instrument more closely, Professor Bose jumped up to prevent him!

Acharya Prafull Chandra Roy, the famous chemist from Calcutta University, also came. He was very popular, a bachelor who gave all his time to his students. We only ever saw him lying on a sofa with his eyes bandaged as he had eye trouble and had come to England for treatment. We went to see him often and he would talk to us lying in that position.

The politicians in our Indian group like Goswami and Chagla used to make constant trips to London to see our leaders who were assembled there. But at that time, I was unable to generate much interest in what was going on in London and devoted myself to study.

Oxford in those days was just recovering from the war. Many of the students, the cream of that generation, had joined the army and had been mowed down. The number of British aristocracy and middle-class, especially the intellectuals, who died in that holocaust was frightening. The "young barbarians" of Matthew Arnold had been made to play a different game and to pay dearly. One felt an emptiness in Oxford despite the semblance of normalcy. One student who had survived was Anthony Eden, later to be Prime Minister. He was then a post-graduate student with Oriental Languages as his subject and he was also a Persian scholar.

One student who did not come back was Raymond Asquith, son of the ex-Prime Minister. He was much talked about. John Buchan paid homage to him in unforgettable words: "Many gave their all for the cause, but few had so much to give. He loved his youth, and his youth has become eternal."

Two famous poets were still to be seen around Oxford. William Butler Yeats lived opposite Balliol College on the High near Carfax. We used to see him taking his daily walk. One of our colleagues, G. K. Chettur, also a burgeoning poet, was his frequent companion. The other was the Poet Laureate, Robert Bridges. Once, while cycling to the Boar's Hill on a sunny summer day with my friend Welingkar, we saw him standing beside his cycle looking pensively at Oxford, probably composing one of his brilliant poems. I have always loved his poetry. His anthology *The Spirit of Man,* which he wrote towards the end of the war, was intended to sustain the spirit of British youth on the battlefield when the outcome of the struggle was in doubt. This anthology is one of my most cherished possessions.

I had to think of the main task for which I had come — the ICS examination. Practically every subject in Arts and Science was included. The maximum mark for a subject was indicated and varied from subject to subject, those taking Mathematics or Classics having the best advantage. Not being either a mathematician or a classical student, I had to have a very large number of subjects to compete for the total of 5,000 marks. Apart from English Language and Literature, Philosophy, History and Economics, I also had to take Logic and Psychology, Sanskrit including the Vedas, and Geography. I joined the Oxford School of Geography which I found very congenial. Professor James Cossar took me under his wing and gave me a lot of his time. I got a Diploma in Geography at the end of the year. I spent a few days of my vacation in his house as a guest. He used to take me out boating on the river and he tried to initiate me into the art of smoking a pipe, without success. I also took up Economics with Professor Nemier. When I came to Logic, I found that what I had learnt in India was too elementary. A. C. Bradley's *Formal Logic* was in quite another league. It was too abstruse and I could make nothing of it. I spoke to Mookherjee of Balliol and he offered to coach me. He came to my apartment from time to time over a month and took me over the whole book, explaining as he went along. What he said was so precise and so clear that I began to enjoy Logic. The examination came in the summer of 1921. I had had barely eight months to prepare. I did not succeed. What encouraged me was that I got the highest marks in Logic and Psychology. I was quite prepared for this failure at the first attempt. For the following year the regulations were changed. The number of subjects was reduced and some new subjects were added. The main innovation was a *Viva Voce,* no doubt, I suppose, to reject whoever the examiners did not find suitable for the Service in the changed conditions after the war. I now had another year to prepare myself.

I became more relaxed after the first failure. I spent the winter vacation in Berlin. I remember the eerie feeling I had as the train approached Berlin in the early hours of the morning. There were no lights on the platform. It was a sad city though totally undamaged. The Kaiser had gone to his exile in the Netherlands. The picture of President von Hindenburg was everywhere. Though his figure appeared to give confidence to the people, the spectre of inflation hung over the population. I had been given an exchange rate of 150 marks to the English pound in London before leaving but by the time I arrived it was 1,500 marks. When I left after a month it had gone down to 150,000. The effect was catastrophic.

However, the old Prussian ways could still be seen. At a dinner party for four people one would still get up and make a formal speech. The food shortage was already acute. I noticed some senior and distinguished waiters who served us take the remains from our plates behind a screen and eat them. When I asked someone who they were, I was told that they were lecturers from the Berlin University! I went round Berlin and saw the tourist sights; the Kaiser's Palace was intact, everything in its place. Charlottenbourg with the *Sans Souci* Palace attracted a lot of visitors. The famed Unten den Linden still had the linden trees. I went to some shows including a Spanish dance and a ballet. I specially remember the ballet in the Eispalast; the sinuous movements of the dancers on ice was a magic world for me. It was at the nightclubs that the Berliners tried to drown their misery and sorrows. I had neither the temperament nor the means to do this. The books of Harold Nicholson, then an Attaché to the British Embassy in Berlin, give a vivid picture. One incident during my stay comes back to me. Being a student of Geography and having an interest in maps, I went to a shop to buy a World Atlas. The shopkeeper would not sell any to me because, I presume, the Indians had fought on the side of the Allies during the war! I did get one in another shop, a marvel of detail, such as only the Germans produce. The interesting thing about that atlas was that, though only recently published, it did not show the new boundaries of Europe, as laid down by the Versailles Treaty.

On my return to Oxford I plunged into work. At the same time I took more interest in life around me. I took up various sports. I found I could still excel in running and jumping. I played tennis for the Indian Club. I joined the College Boat Club and was given trials on the Fours — I gave it up after one term as I could not bear the cold in the crew's scanty costume. There was a large contingent of Indian students and we organized a reasonable social life. There was the Oxford Mujlis parallel to the Oxford Union which met once a week and held debates in the same style as the Union. For one term, Goswami was the President, Chagla the Secretary, and I was the Treasurer. I used to be rather shy and self-conscious and do not remember having taken any active part in these debates. The shining lights were the other two. K. P. S. Menon was a very good speaker, one of the best among us, with a sly wit which was most engaging. In the early days there used to be evening parties where K.P.S. would soon be the happiest merrymaker in the place.

I pursued my studies in Economics with Professor Nemier. I sat for a Diploma in Economics. Just before the examination, I had read a series of articles in a magazine on the losses suffered in merchant shipping by the nations at war. I had a memory for figures and I remembered all the hundreds of them given in the articles. In the examination, I was able to use all the figures. Professor Nemier was one of the examiners. I got an Honour to my credit and later he expressed surprise that I could quote so many figures correctly! The ICS examination came. This time I was successful. I was delighted and immediately sent a telegram to my father. I knew it would make him happy. At last, I had achieved my life-time ambition. Among the others were K.P.S. and N. R. Pillai, who distinguished themselves in the Service later. Another was Sukumar Sen. Our families came from the same village in Vikrampur, Dacca. On our return to India, we made it a point to visit our village together. Never had such a thing happened to a poor village in the past — two ICSs at the same time!

I gave myself up to all the pleasures of Oxford in the third and last year. Punting in the Cherwell with Welingkar for company was a real pleasure. We used to tie up the punt to one of the willow trees which lined the banks of the Cherwell, and lie lazily reading or dreaming. Sometimes we used to row close to one of the locks and find a haystack, lie against it and read poetry. The Eights Week — the annual summer boat races — was a colourful affair. Pretty girls, sisters and relations of boys at the University, used to join us for the day to enjoy themselves and look for suitable partners for life!

In the winter of 1921 I visited Berlin again, this time in company with two other students. One of the two, Pillai by name, I did not know very well. Soon after we arrived in Berlin, I discovered that Pillai had some unsuspected political affiliations. One day he took us to a darkly lit office with a somewhat sinister atmosphere which must have been an Intelligence Centre. On New Year's Eve, he took us to some secret clubs which looked dark and inconspicuous from the outside but were brightly lit inside with people milling around drinking and talking. For the first time in my life I took a drink at every place we went to that night. When the night was spent and we had to go home, I had to be helped back to the hotel by my more experienced companions. This, I may say, was the only time in my life I have found myself in that condition. However, whatever Pillai was up to, or intended to do, I and my other companion remained outside the circle and returned unscathed.

After one year's probation as a member of the ICS, my three-year period in Oxford came to an end. As the time approached for me to leave, despite my nostalgia for home and my people, I felt a great sadness. The three years had been among the happiest of my life. I had had my anxieties, specially during the first two years, but even then I had a feeling of happiness I had not experienced before. What is it that Oxford gives to those who enter its hallowed portals? One may read in English anthologies the tributes paid by many writers of prose and poetry to this ancient place of learning. To me, also, coming from a very restricted life, Oxford meant something special. The quality of Oxford is that by the manner of life there and the character of her institutions she helps to heighten those aspects of youth that are most precious. She brings a freedom of spirit to think and reflect and to indulge in impossible dreams. I have visited Oxford in later years but I could not even faintly recapture the halcyon spirit of those student days. For that one would have to be young again!

Chapter 2

In the Indian Civil Service

An ICS in the making

RETURNING to my country in October 1922, I spent a few weeks with my parents in Dacca. Then I assumed my post as Assistant Magistrate in Mymensingh (now in Bangladesh), a couple of hours' journey by train from Dacca, making it possible to take weekend trips home.

The official atmosphere in a district in those days was set by the District Magistrate, the most senior ICS officer. His word was law for all. The atmosphere I encountered in my first posting was far from friendly. The men in the ICS were still unaccustomed to having an Indian by their side. The District Magistrate was correct in his official dealings but kept me at arm's length in social life. My lodging was what might have been an old barracks converted to living quarters. My suite of two rooms was at one end; the other end was occupied by the District Club which was exclusively for Europeans. Apart from the ICS officials, members included other people — Greeks, Armenians — who were in the jute trade. The district was one of the most important jute-producing centres. I naturally and quite innocently asked for membership of the Club because I liked to play tennis. I was found unacceptable and black-balled! I was furious but this was the way the British behaved in India in those days. They had not yet realized that their "Empire" was at an end. This was, no doubt, a reflection of the attitude of the District Magistrate because, when he left after a few months and went on home leave, his successor came along to me to express his regret and ask me to join.

This was a period of apprenticeship in my work and at the same time an initiation into the esoteric freemasonry of the ICS. I was a young man of twenty-four. Apart from having to learn the work of the various branches of administration, I was invested with judicial powers to try criminal cases. I began with third-class powers and rapidly went on to second- and first-class powers which widened the jurisdiction and increased the maximum sentence I could impose. I thought my promotion was more rapid than my performance justified. I remember once, while cycling from the house to the Court, I had already written out all the right arguments to release a man I had convicted in the judgment. I came home and acted on that intuition! From the very beginning I took a dislike to judicial work, specially having to take down the deposition of witnesses in longhand. I found it too boring. I carried this dislike with me when the time came to make a choice for my future career.

The Service was divided into two branches — the Executive and the Judicial. All members of the Service were expected to experience the responsibilities of both before finally adopting one branch or the other. Those who preferred a regular, peaceful, independent life went for the Judiciary, though its career possibilities were very much more limited than those of the Executive. But sometimes this was not a matter of individual choice; the superior officers at the helm of Government had a say. One who proved himself temperamentally or otherwise unfit for Executive charge was pushed into the Judiciary where he might be less embarrassing. The majority of Indians in the Service found themselves in the Judiciary, either by choice or by compulsion. This, of course, by no means implied that men in the Executive were a select band of superior men. Far from it. There were examples of misfits who did not raise the prestige of the ICS but who became a source of embarrassment to all. When my time came to be a District and Sessions Judge for a time, I was firm in expressing the wish to remain in the Executive even though it meant temporary loss of the upper grade that one naturally sought. I stayed in the Executive throughout my ICS time.

Towards the end of the first year as Assistant Magistrate I had to join a Settlement Training Camp to learn all about the complicated land tenure system in Bengal, which was a legacy of the early British administration under the East India Company. We lived under canvas for three months and had to wade through water sometimes waist-deep drawing the measuring chains to demarcate any change in the boundaries of land holdings that might have occurred since the last settlement, a decade ago. It was good discipline, if not good instruction. But I was glad to get back to normal life.

I spent the next five years taking charge of sub-divisions — Midnapore, Rampurhat and Serampore. Sub-divisional charge was the lowest rung of the administrative ladder but one which gave the greatest opportunity for officers to find out the problems of daily life of the people, especially those of the poorer people in the rural areas. In truth one lived with the people with all their problems; even their private life was exposed. From the point of view of learning about the people and the country, these five years in the sub-divisions were the most satisfying. I was the master of the sub-division and there was very little interference from the District Officer whom I went up to see from time to time to discuss any problems of administration that we might have had. There were no politics reaching down to the sub-divisions in those days. My powers as a Magistrate to try criminal cases also reinforced my position. I had to do everything — manage the various services of the administration, try criminal cases, cope with cholera and other epidemics. I remember going from house to house in a village to have cholera cases removed to the district hospital and making sure that cholera inoculation was available to all who needed it. Sometimes people came to me with purely domestic problems for my advice. A Scottish lady once came to seek my advice on whether she should divorce her husband for his drunkenness! I look back on my sub-divisional days as a rewarding period of my official life. As I went higher up the ladder, as District Magistrate, Divisional Commissioner, Secretary to the Provincial Government at Headquarters and finally to the Central Administration in New Delhi, I felt that I had gone further and further away from the people and their real problems.

There was a break in my sub-divisional career in the winter of 1926 when I was called to the Secretariat in Calcutta to write the General Administration Report for 1925. I found the more senior officers in the Political Department where I had to work somewhat uncooperative. A senior Indian officer, particularly, did not appreciate another Indian officer so much his junior poking his nose into his affairs, as I had to collect material from everyone for writing the Report. However, this interlude was unimportant except that it gave me an opportunity to get an insight into the mysteries of the Secretariat which was the apex of the administration.

For the next two years I went back to sub-divisional duties. When I was in charge of Serampore sub-division, there was a temporary vacancy in the post of the District Magistrate. I was asked to take over being given the position over the head of a very senior man who was next to the District Magistrate. It was unjust as it appeared to me that his only fault was that he was not in the ICS! During this period I had a serious communal disturbance to face — widespread riots between Hindus and Moslems. I concentrated all the police I had in the district in that small affected area, cordoned off the mosques with police, to protect them from violence as well as to prevent the mosques being made into a base for organizing more violence. I moved round the affected area all through the day — somewhat unusual for a District Magistrate to do. There was not a single incident.

As I emerged from my sub-divisional days I realized that I was developing certain characteristics as an ICS officer. While taking great personal interest in the problems connected with the daily life of the people, I had developed a habit of keeping myself detached. This was probably no more than my self-consciousness and shyness requiring protection in the form of detachment. It was a sort of mask that I wore but what people saw in the mask became my true character to them. The consequence of this was that people around me never tried to be familiar with me. I was told that they feared me and found me somewhat unapproachable.

The first stage of my initiation was now over and I was experienced enough to take more responsible duties as a District Magistrate. I was now ready to penetrate into the deeper mysteries of the ICS.

As a District Officer

MY FIRST POSTING to a district was to Barisal at the end of 1928, as Number Three in the executive hierarchy. Soon afterwards, I took over as Number Two when the District Magistrate resigned as a result of threats by terrorists for his speeches in the Central Legislative Assembly in New Delhi against the terrorist movement in Bengal. Immediately I was exposed to the political forces working at that time.

Agitation against the British rule in India had, from the very beginning, taken two parallel forms — constitutional and revolutionary. The constitutional agitation was directed towards Home Rule and was for securing a greater number of places in the Imperial Services, civil and military, for Indians. The reluctance of the British Government to yield any ground had led to a hardening of the attitude of the Indians who made progressively greater demands. At the same time, distrust of British motives had given rise to a revolutionary movement demanding full freedom from British rule in India. On several occasions, the two movements had merged together, but not at the higher levels. The leaders of non-violence and constitutionalism had consistently and faithfully maintained their basic principles against the revolutionary pressures. The emergence of Tilak and Aurobindo and others had increased the pressure on the British. The Morley-Minto Reforms of 1909, designed to meet the situation, had done little to satisfy the political aspirations of either the Constitutionalists or the Revolutionaries. The enlargement of the Governor General's Legislative Council and the Provincial Legislative Councils did not work, as these Councils were found to be no more than advisory bodies and did not provide any trace of parliamentary democracy or ministerial responsibility. The Revolutionary movement in India found an ally in the *Ghadr* movement which had originated in North America and now, under its leader, Lala Hardaval, established contact with Germany, then about to become engaged in a world war with Britain.

The intervention of World War One strengthened the hands of those who wanted India to be a self-governing dominion. At the end of the war the Montagu-Chelmsford Reforms were introduced (1918) with the vague announcement by the Secretary of State, E. S. Montagu, that "the policy is that of increasing association of Indians in every branch of administration and the gradual development of self-governing institutions", which did not satisfy anyone. These halting steps brought Mahatma Gandhi onto the political scene with his Passive Resistance movement. The Indian masses came into open political conflict with the British rule. The Punjab rebellion in 1919, culminating in the Jallianwalla Bagh massacre, further exacerbated the politically tense atmosphere.

The Government of India Act of 1919, based on the Montagu-Chelmsford report, introduced "dyarchy" in provincial administration, which was firmly resisted by leaders like C. R. Das in Bengal and was brought to a standstill. While the non-cooperation movement of Mahatma Gandhi continued to expand, the British Government tried various concessions to quieten the situation. The Lee Commission (1923) proposed some positive recommendations on the composition of the ICS and Indian Police Service, in the context of Lloyd George's *Steel Frame* speech (1922). These developments constituted a small step forward. It may be mentioned here that it was only after the great sacrifices that India had made in men and money during the war that, in 1918, Indians were regarded as fit to get the King's Commission! Indian public opinion, however, did not view all these changes as being more than cosmetic. Further, the Indian political leaders no longer placed the same emphasis on these smaller questions as they had done in the past. Their sights were now set on total change.

The Bengal Revolutionary movement had gathered strength and was involved in several political murders, especially of police officers and persons assisting them. The Bengal Criminal Law Amendment Act of 1925 put many of the suspected terrorists in internment camps. But the movement spread to other provinces. The arrival of Lord Irwin as Viceroy, who had sympathy for the Indian cause and at whose persuasion the Simon Commission was set up to enquire into the working of the 1919 Constitution, was an opportunity for him to announce (October 31 1929) that "the natural issue of India's constitutional progress is the attainment of Dominion status". This statement was made after publication of the Nehru (Motilal Nehru) report asking for the immediate granting of "full responsible government on the model of the constitutions of the self-governing dominions".

This was the background to the political situation when I came to Barisal as one of its executive officers. It was one of the most important districts in that part of the country, with a large population, predominantly Moslems. Jute was produced in the area and business was done with a number of British and other European traders. It was also a centre for the British Steamship companies who operated all along the Brahmaputra from Assam to East Bengal. The district Club was a busy one and the atmosphere was much different to what I experienced in Mymensingh. I was on my own and threw myself into work and play.

One of my first official duties was to try a Sedition case against a band of young men alleged to belong to the Revolutionary movement and alleged to have been engaged in terrorist activities. The case went on for months. I do not quite recollect the outcome but it brought me right into the public eye. The sympathies of the people were naturally with the young men who were regarded as sacrificing their lives for their country. My father, travelling by steamer from Dacca to Barisal, told me, rather hesitatingly, that he overheard a group of men on the steamer discussing newspaper reports about the trial and making very disparaging remarks about the young magistrate. He didn't want me to do anything or change my attitude to the case in any way unless I was myself convinced that it was the right thing to do. That was the kind of relationship I had with him. On one occasion, the young men on trial had to be sent back by force to the jail for contempt of Court. I was warned by the police that this might endanger my personal safety.

About the same time, I was arranging an Exhibition to promote the various products of the district. The Exhibition had no political significance, yet I received threats and was told that the Revolutionary party was against it and I was advised to abandon my plan. I ignored the threats. On the opening day, some youths and a few elderly men lay on the ground before me and blocked the entrance. With me were the Superintendent of Police (a senior Englishman) and the Kotwali Officer (in charge of the police station of Barisal). As I stepped carefully over the bodies the other two followed me. Unfortunately, the Englishman had a dog with him who also followed us in. This was a gross indignity to the demonstrators and they swore revenge. Within a few days the Kotwali Officer was stabbed to death. The Englishman had a nervous breakdown and took off in his steam launch (we each had a launch instead of a staff car — being the only means of quick transport for us in that river district), applied for long leave and did not return to the district. Every afternoon or evening I used to cycle to the Club. I received dire threats that I would meet the fate of the Kotwali Officer or have acid thrown in my face. I didn't take any notice remembering my mother's oft-repeated words, "whatever you might do, you would not be able to avoid what is written by Providence on your forehead!"

Apart from this political embroilment, I had another urgent problem to attend to under orders from the District Magistrate. We received reports of serious riots in the interior between Moslems and Hindus with indications of numerous casualties. It was a night's journey to the area by steam launch. I was accompanied by about fifty armed police under the Superintendent of Police. Before we reached our destination I explained to the police in what circumstances they could open fire. I was sure that firing would have to be resorted to. As we approached the area I saw crowds standing on either bank calling for protection. At one spot some villagers brought news that they were under attack and needed urgent help. I asked the Superintendent of Police to disembark with all his men except two to protect me from attack. They soon disappeared from sight. When a spot was found to tie up the launch I disembarked. A man came running up to me to say that several rioters had entered his village and were bent on killing and raping his people. This was a traumatic moment for me. I had no police with me except my two personal guards. When I arrived at the spot I found no trace of any rioters. Apparently, it was mass panic at work. In the evening when we all gathered together the story unravelled itself. A young Hindu girl had been abducted and immediately given in marriage to a Moslem youth according to Islamic law. This then led to riots between the Hindu and Moslem communities and some lives had been lost. I asked for the new bridegroom to be produced. The Moslem elders of the village denied all knowledge of him. He was, however, produced under threat of punishment. I despatched my launch to bring the newly wed bridegroom from the sub-divisional jail where he had been detained for another offence. It was after midnight when he was brought before me. An assembly of all the villagers from both communities was called. Before them, a *talaq* (divorce) was performed according to Islamic law. The young innocent girl — she could not have been more than twelve years old — was then handed over to her family with a stern warning that the family or their friends in the village must see to it that this incident did not, in future, harm her socially or militate against her in any way. The dawn came. The journey back, after the sleepless nights and all that I had gone through, was a pleasant relaxation. Ten years later a friend of ours came from that district and said that they still talked about the young ICS officer and his original exercise in restoring communal harmony!

While in that district, it was my painful duty to supervise the hanging of a young man for murder. The District and Sessions Judge later told me that he had sentenced him to hang because in that way his sentence would have to go before the High Court for confirmation and the High Court, on the kind of evidence presented by the prosecution, was sure to set aside his judgment. (If he had found him guilty of manslaughter instead, as he said he should have done, he would have had to sentence him to a long term of imprisonment which he was reluctant to do.) The sentence was not set aside and I saw the condemned man every day for about a week before his hanging. He was young and looked very much like anybody else. He had a quarrel with his mother-in-law because she took his wife back to the parental home and in a fit of temper he hit her with a kitchen knife and killed her. It was a very sad case and caused me a great deal of worry.

My next posting was as District Officer of Noakhali (now in Bangladesh). Before taking up my position I spent a few weeks in Darjeeling where the Bengal Government had moved for the summer season. Among the Members of the Executive Council I met a very colourful figure, Alhadj Sir Abdel Kerim Ghaznavi, who held the portfolio of Irrigation. He told me that Noakhali had one big problem to which I must give immediate attention. The foundations of the town were being seriously eroded by the action of the Noakhali river and the only way to stop this process was to construct an earthen dam to stop the flow of the river along the town. I was a complete ignoramus in such matters but, since the advice came from the highest authority of the land, I took the matter to heart. As soon as I arrived in Noakhali, I tackled this problem. I found that my immediate superior officer, the Divisional Commissioner, was totally against it, but then I also discovered that there was some personal family quarrel between the two. In the confusion, I decided to follow the advice of the Member of the Executive Council who, after all, had at his disposal the best irrigation experts in the country.

I arranged that the dam was to be constructed during the winter months when the flow in the river would be at its lowest. Thousands of labourers were collected from around the town. Strong protest deputations came from villages located on the upper reaches of the river especially from Maijdi which was an important centre for recruitment of sailors for foreign merchant ships. In my youthful enthusiasm I boldly proceeded, overriding all objections, even of my Commissioner.

The work started. Booths were set up to supply food and water to the labourers so that they would be in a position to work long hours. The construction started from both banks but, as the gap for the flow of water became smaller and smaller, the force of the river water became all the greater. I was called in by the engineers as the labourers were getting exhausted and could not be held without my personal presence and admonition. I remember I stood on the site for nearly twenty-four hours, all through the day and the night, cajoling and encouraging, till the gap was finally closed. When it was finished I was utterly exhausted.

Within a few days the villages in the upper reaches of the river were flooded by back-up water from the dam. There were angry protests. When I visited Maijdi and saw the conditions I wondered if I had acted correctly. But before I could do anything, people from some of the affected villages cut the dam open in the dead of night and solved my problem of conscience! This ending to the saga greatly pleased my Commissioner and our good relations were restored. Nor did I lose the goodwill of Sir Abdel Kerim for he believed that I had followed his directive faithfully. He often asked me to his house to show me the many addresses and citations he had received for his benevolent work in different parts of the country which were carefully preserved in silver caskets. He was a simple man. Later I learnt that he was not always of sound mind. In retrospect, I consider the dam at Noakhali to have been the greatest folly of my life!

The Lahore Congress of 1929 adopted a resolution demanding complete independence and authorising a campaign of Civil Disobedience in support of that demand. Mahatma Gandhi's Dandi March in April 1930 caught the imagination and touched the patriotism of the entire nation. In my district, the Congress workers decided to try and close down all Government offices for a day by lying down in front of the doors and impeding access. I issued orders that the offices should remain open and that the staff should attend. I had to give the lead and with some difficulty reached my own office. This made me very unpopular. There were protest marches. An article in the local newspaper attacked Government authority. The editor was prosecuted for sedition. My judgment, sentencing him, was given headline publicity all over Bengal. I often wondered if I was doing right in all this. But I was a confirmed constitutionalist, and our leaders were still trying to proceed constitutionally (Motilal Nehru report 1928).

Some time later when I was just recovering from these experiences an urgent message came from my Commissioner in Chittagong: "Chittagong Armoury looted by terrorists. Expect similar rising in Noakhali. Send all available armed police force to Chittagong immediately." Noakhali was a comparatively small district and the police force was small. However, orders had to be obeyed. I immediately asked all the officers in the town to assemble at the Treasury building in the old town to which the police armoury was attached. I sent the major part of the police force to Chittagong keeping a minimum to guard the armoury and the Treasury. We sat through the night expecting an attack at any moment. But the night passed peacefully and there was no sign of any terrorist activity. Later the details of the Chittagong armoury-raid came through. About a hundred youths under a much-respected leader, Surya Sen, a teacher, dressed out in uniforms, had attacked and looted the armoury, and also took over the telegraph office and the European Club and got away with the guns and rifles. This was a most daring exploit and created a great stir. Among the public there was much admiration for their courage and achievements. A few days later as I was getting ready for the office, I received a message from my Sub-divisional Officer at Feni, a distance of thirty or forty miles from Noakhali, to say that some of the escaping Chittagong raiders had shot a police officer who was searching a train at the local station and I was needed urgently. I immediately motored down to Feni but there was no trace of the raiders. Some time afterwards, there was a shoot-out between Sir Charles Tegart, the formidable Police Commissioner of Calcutta, and the raiders who had gathered at Chandanagar, near Calcutta. Some of them died and some were arrested.

One of the young men who was involved in the original raid on the armoury at Chittagong was a first cousin of mine who died from burns he received in the attack. It was a hectic and very unsettling time.

My next posting was to the Secretariat at Calcutta to take up an entirely different kind of work. I entered the Bengal Secretariat as Deputy Secretary in the Political Department at the beginning of 1932. I had just got married and a party was given for us by an Englishman who had been very frosty to me when I first joined the Service. He was now the Chief Secretary, a W. S. Hopkyns. I appreciated the gesture and it showed how times had changed.

I took over from a British officer whom I had known before and who was a bright, competent man. His table and the floor were littered with files. He said he had found it almost impossible to do all the work that had been required of him. He was kept awake all night by the Central Telegraph Office asking for clearance of press telegrams as, in addition to his other duties, he was also the Press Censor! He went on home leave to recover. In those days, the established Secretariat practice was to record each and every remark or observation on any official matter on the file; there were no committees or individual discussions to resolve problems. The most efficient Secretariat officer was one who could express himself best in his notes. The number of files that came daily to one's desk was enormous. I decided to introduce some rationalisation into the filing system in my new office. Files were to be divided into three categories — those to be initialled without reference to the notes, those in which the notes had to be read before initialling; and those which were to be treated more seriously and, if necessary, taken home for further study before disposal. In a few days, I cleared a mountain of files from the desk and from around my feet. As for the Telegraph Office, I gave orders that I was not to be disturbed after midnight except in cases of real emergency. This helped a good deal except on certain special occasions such as during a Congress session when I was available throughout the night. I also made it a point to leave the office by the appointed office-closing hour each evening and I went for a game of tennis and relaxed. It worked.

The political atmosphere in those days was supercharged with negotiations and discussions. The Simon Commission report (June 1930) made no concessions to Indian public opinion as it made no mention of Dominion Status or of a truly parliamentary system of government. The first Round Table Conference in London (November 1930 - June 1931) was boycotted by the Congress. The Gandhi-Irwin pact (March 1931) created a more favourable atmosphere and Gandhi decided to attend the second Round Table Conference (September - December 1931). Arising out of the discussions, several Committees — Lothian (Franchise),* Percy (States), and Davidson (Finance) — visited India in 1932.

*R. A. (Rab) Butler, whom I have quoted in my Preface, came as a member of the Lothian Committee to Calcutta. Being then in charge of the Press in Bengal, I invited him to my home for tea (in those days 'cocktail' parties were unknown and the British institution of tea parties remained supreme). As he was about to leave he asked me "What can I do for you?" The egoism of the young man, a little younger than myself,

While these constitutional discussions were going on, the terrorist situation in Bengal had assumed serious proportions. In 1932, when the Bengal Secretariat, with the senior officers, went up to Darjeeling for the summer, I, as Deputy Secretary, had to decode all the official telegrams for the Chief Secretary. Week after week I had to bring messages to him about the murder of British ICS officers, three in one district alone (Midnapore).

When we came back to Calcutta I was asked to take all possible means to prevent the Press from publishing news or views which could give encouragement or comfort to the terrorist organizations. Very often newspapers published comments which in more normal circumstances would have been quite harmless. For instance, when some of the political leaders who were following a policy of non-violence said that terrorist activities were due to the uncooperative attitude of the British Government in meeting constitutional demands, this was no doubt perfectly true. But in the situation then prevailing, such words from political leaders could only encourage the terrorists. I was given a free hand to devise measures to prevent the Press from giving succour to the terrorists. My first step was to draft legislation empowering the Government to demand security deposits of specified sums from newspapers against publishing material encouraging or supporting terrorist acts. This was issued as an Ordinance. I was to operate it. The issue of the Ordinance brought about an instant change. The newspapers were often made to publish offending material under duress and the Ordinance made it possible for them to resist yielding to such pressures. I established close personal contact with all the important newspapers. I told them that if I found them cooperative, I would protect them instead of penalizing them for occasional lapses. I had a list of all the newspapers and their telephone numbers in front of me. Whenever any matter which was likely to create difficulty came to my notice — a press telegram, a statement in criminal proceedings, a news item — I would take up the list and telephone my instructions and advice. This was acceptable to them because of the personal understanding I had established. As far as I recollect I did not have to ask any newspaper to make a security deposit under the Ordinance all through the three years I was there, (1933-35). On one occasion a girl who was being tried for attempting to shoot the Governor of Bengal at the annual University Convocation ceremony, made a fiery statement before the Court saying that what she had done was for her country and she vowed she would do it again if the opportunity arose. As soon as this came to my notice I rang up all the newspapers advising them not to publish the statement. I said that the accused's statement in Court was privileged but its publication in a

shocked me. We have had no occasion to meet again. But when I came across his remarks about "egoism" as essential to autobiography, I remembered his egoism, now nearly half a century ago.

newspaper was not and if they published it they would be liable to the due process of the law. The Bengal Legislative Council was then in session. A member moved an "Adjournment Motion" to raise this matter and condemn Government for its action in muzzling the press. (As the Council under the 1919 Act was not a wholly elected parliamentary body with 'ministerial responsibility', there was no such thing as a "No Confidence" motion.) There were several official members, myself included, occupying the "Treasury Bench". The extraordinary aspect of this debate was that the Executive Council Member in charge of Law and Order, Sir William Prentice, replied defending my action, yet he never asked why I had done it though I sat next to his room in the Secretariat. This was one supreme example of that characteristic of the ICS of allowing officers to discharge their responsibility to the best of their judgment and defending them to the last, trusting they had done it in good faith and for good reason.

Indeed, I wish to pay a tribute to Prentice who was my hero. He had a formidable reputation for efficiency and forthrightness but he was always just and fair. His sudden death left a big gap in the Bengal administration. In his last will he had stipulated that his body should be cremated and his ashes scattered over the waters of the sea in the Bay of Bengal.

I was a member of the Bengal Legislative Council and, indeed, the youngest member in the House. It was a stimulating experience. I sat behind Sir William Prentice and my principal job was to reply to written questions when asked to do so. I hardly took any part in the debates. The great figure in the Council was Shyamaprasad Mookherjee, one of the most brilliant men of our times and an orator of high order. I had a lot of contact with him later on. One of the elected members who took me under his wing was N. K. Basu, an advocate of the Calcutta High Court, whose brother was in the ICS and was much senior to me. Basu was a fine speaker, always well briefed on a subject. On the Government side, Sir Bijoy Prasad Singh Roy, who was a great gentleman and a scion of an old Bengal family, enjoyed a high reputation. I had the privilege of working with him later as Revenue Secretary to the Government when he was the Minister in charge.

By this time Mahatma Gandhi had returned from the second Round Table Conference and was disappointed at the failure to settle the issue of communal representation. Soon after, he was arrested for starting the Civil Disobedience Movement. The generally disturbed political atmosphere once again provided fertile ground for terrorist activities. It was a difficult time which made my task in my own little world a constant battle of conscience. I was right in the middle of the political cross-currents and all I could do was to carry on with as much fairness as possible. In those years I acquired a great deal of experience and it was also a testing time for one with my kind of upbringing and experience.

It is true that this working relationship between the Government and the Press in Bengal at that critical time did not find favour with the political leaders (see Nehru's *Discovery of India*). But if the purpose and the limits of this cooperation had been kept clearly in view, the reaction might have been a little less critical! I went on long leave at the beginning of 1935.

Happy Days at Maldah

WHEN I WAS growing-up, the Indian Civil Service had a prestige quite without parallel. It was a career prized by the most ambitious young people. But the winds of change had begun to affect the attitudes of ordinary people to the Service. My own feelings in this situation (worked out during my long leave abroad) were that whatever political changes might come about in India the Government of the day would need a strong, disciplined and efficient Civil Service. That the Service was largely British in composition did not make us less Indian than others. True, we did acquire some characteristics which distinguished us in the eyes of our countrymen but those were superficial. At heart we were the same. How then, in the prevailing political turmoil, should we conduct ourselves and be useful to the country? None of us were in the inner conclave of policy-makers which looked after British interests in trade, defence, etc. We were engaged largely in matters which were of direct concern to the people.

On my return, I found the general political situation much quieter. The 1935 Act had abolished the old dyarchical system of government and introduced provincial autonomy. The Congress now appeared more receptive but waited till the Viceregal announcement that the intention was to strive for "full and final establishment . . . of parliamentary government" to contest the elections. The Congress secured majorities in almost all provinces except Bengal and the Punjab where the Muslims won a majority.

I was posted as District Officer of Maldah. It was a posting much sought after by the British members of the Service. For one thing, it was somewhat in the political backwaters of the country and one could work in peace and quiet. Another was that it was a haven for game birds for those interested in shooting. After my stormy time in Calcutta I very much enjoyed coming to the peace and quiet of this district. One of my immediate predecessors was James Peddy, a man of vision and understanding whose name was revered in the remotest villages. He had a true missionary spirit and often disregarded all personal considerations of comfort and safety to bring succour to a cholera-infested area and would spend nights in local village huts with the people. I made him my ideal District Officer and sought to emulate him. (Sadly, on his transfer to

Midnapore in 1932 while visiting an exhibition in the collegiate school he was assassinated.) I also tried to revive the spirit of my sub-divisional days in that I sought to make the life of the people my own.

The roads in the district were impassable except in summer. So, no doubt at the initiative of some District Officer in the past, the Government had thoughtfully provided in the adjoining garage, not a motor car but an elephant! My elephant's name was Rampyari. I travelled extensively by elephant through the jungles and impassable roads in the interior. Every afternoon Rampyari was taken into the bush by her driver to collect a huge load of food. She was a self-server! My relationship with Rampyari helped to rekindle a special fondness for elephants I had as a boy especially having read the books of "Elephant Bill" who was a forest officer in Burma and who was well-known for his intimate and authentic elephant stories. One of my favourite stories was the one in which a herd of elephants trained in logging was being withdrawn from the Burmese war-zone to avoid their falling into enemy hands. The herd neared the Indian border led by a big tusker. A narrow ledge on the side of a mountain had to be negotiated. The tusker halted and seemed to ponder for a moment or two to see if it was at all possible to face the extremely narrow ledge. He realized that the risk had to be taken and led the way into the danger. When the last elephant in the herd had crossed the tusker fell down dead overcome by the effects of the awesome risk he had taken! Another story was about the way an expectant mother elephant was protected by another female elephant (the equivalent of an aunt) from being disturbed by anyone, man or beast. These were beautiful stories and now, years later, I had an elephant of my own! Indeed, those earlier feelings gave me an insight into the ways of the elephant.

I had personal experience of the staunch courage of elephants. We were out on a leopard hunt. I was mounted on a big elephant who was the leader, an animal who had had a lot of experience of hunting. The leopard was sighted and from the elephant's back I opened fire. It was only after the shoot was over that I noticed blood streaming down the ear-flap of the elephant. One of my bullets had gone right through it but the pain had not caused him to panic!

The people of Maldah received me warmly and responded to my attempts to revive, enlarge or rebuild the local institutions. I gave particular attention to improving the facilities of the District Hospital, the Sericulture Training Institute (this being an important centre for growing silkworms and for weaving silk cloths), and to the building of a new power station. Maldah's fame came largely from Gaur and Pandua, the ancient capitals of Hindu and Pathan kings. Precious statues of Hindu gods and goddesses were strewn all over the place. It seemed strange that no one had thought of collecting and preserving them. I started a district-wide hunt offering some suitable incentives! Within a short time we had a large enough collection to build a museum. The Maldah

Museum still stands and houses a very valuable collection dating back to the fifth century. While I was there experts from the Archaeological Survey of India came to inspect the collection with a view to removing the valuable pieces to the Calcutta Museum. However, I insisted that they could only have duplicates. I wanted the importance of the district as the home of the ancient Hindu kings to be recognized through the collections exhibited in this museum. The architectural ruins of the Pathan kings have been made known to the world through the paintings of Daniel. These are impressive ruins. The proportions of the architecture are delicate like a poem. The stone carvings are exquisite. Maldah should be a tourist centre, if not for foreigners, at least for Indians who want to know their own history.

Maldah is also a district famous for its game. The many swamps are a favourite summer resort and breeding place for birds migrating long distances. Once, to my great dismay, I brought down a high-flying bird which turned out to be a Siberian duck come all that way to breed. It had long been the practice of District Officers to spend the winter months under canvas, moving from village to village to establish contact with the rural people, to hear and redress their grievances, and at the same time to do some shooting. I fell into the same practice though shooting and killing game was not one of my favourite pastimes.

Sir John Anderson (later to be Lord Waverly) was then Governor of Bengal. He was reputed to be one of the most able civil servants ever produced by Britain. Stories of his exploits were in wide circulation. He had been sent to Bengal to bring the terrorist movement under control. (Later, relinquishing charge as Governor he was to become Lord President of the Council and was to assist Prime Minister Winston Churchill in the conduct of the war.) I came to know him when I was Press Officer in Calcutta. I invited him to visit the district. He agreed to come and act in his formal role but he asked me to organize a partridge shoot!

I sent out my experts to mark out the best partridge locations and laid on all arrangements. A few days before the Governor was due to arrive all the partridges in those selected places suddenly scattered. What was I to do? I was told that live partridges were always available in the Calcutta market. I immediately sent one of my officers to the market to buy as many partridges as he could. The partridges arrived and were released the day before the shoot was to take place. On the day, we were all mounted on elephants and advanced in line. The rule of the shoot was that a partridge could be shot only when it rose right in front of one. My first *faux pas* was that, following that rule, I was the first to bring down a bird. But this was strictly against protocol since the chief guest must have the first shot! When the shoot was over all the partridges that were brought down were laid out in a row before the shooting party for inspection and mutual congratulations. Suddenly, much to my embarassment, Sir John began to enquire how it was that the partridges had a dark shade on the neck which, according to the book he had read in the train from Calcutta, only partridges from upper India should have! Knowing him, I thought my best course was to maintain a thoughtful silence. However, when he was due to leave I was assured by his Private Secretary that he was pleased with his stay and was taking away a very good impression of my work as a District Officer!

I had no real interest in shooting. It occurred to me that there should be a Convention regulating shooting. With some difficulty, as I had very little reference material available, I drew up a draft Convention and sent it out to the proper authorities for comment. But before I could proceed further, I received orders of transfer. My next posting was to be Midnapore, the most disturbed and the most difficult district in Bengal. Three British District Officers had been murdered there. Two other British officers had been posted to that district but it was still out of control. I do not know if it was my term as Press Officer which impressed Sir John or if it was his visit to Maldah and his partridge shoot but I was told that I had been selected by him personally to try and "pacify the district".

I enjoyed my term in Maldah and I left with regret. One of my farewell presents was an album of photographs of Gaur and Pandua. In a quiet or reminiscent mood, I sometimes turn the pages to recall those happy days.

Midnapore

I LEFT the train at Kharagpur, the big Railway Centre, to take the road to Midnapore. As I opened the door of my compartment I was met by a troop of Gurkhas standing with fixed bayonets. They surrounded me, protectively, and escorted me to the motor car. It was a shield of bodies to prevent an assassination attempt! My car was parked between two trucks both filled with armed police. Two personal bodyguards took their place inside the car, one beside me and the other in the front seat and both wore loaded pistols on their belts. The military procession got underway, crossed the Cossye river, and finally brought me to my destination.

My residence was an old house dating back to the last century and was built by indigo planters for whom this district of early British settlement was a most important centre. Armed sentries were posted on all sides. As evening came, searchlights were switched on to light up the approaches to the house. This was Midnapore!

A day or two after my arrival, I was invited to distribute prizes at the finals of a Football Tournament. The entire playing-field was surrounded by armed police. (It was on this ground that one of my predecessors, Burge, had been shot while actually playing in a game.) There was a sprinkling of spectators sitting round the field. The match started at 5 p.m. At half-time I noticed the spectators getting up in a hurry to depart. I learnt that this was because of the curfew which had been imposed on the town since the incidents in 1932 — a curfew from 6 p.m. to 6 a.m. It had remained a way of life since then.

About this time also came the anniversary of Peddy's death. Prayers were offered at the local church. Near the entrance to the church were situated the three graves bearing a grim testimony to the troubles.

In a few days I realized the kind of district I had come to. Unlike the normal practice today of briefing and debriefing, I carried no instructions, written or verbal, from the Government about the situation in the district and what I was expected to do. I had come with a fresh, or rather blank, mind. The idea was conveyed to me second-hand that the Governor, Sir John Anderson, had specially selected me to "pacify" the district. I had known of the terrorist activities and in a general way followed the sequence of events there. This was how administration was run in those days — every officer had to learn to be self-reliant. Now that an experiment was to be made of giving responsibility to an Indian officer, why burden him with instructions? That must have been the philosophy.

Within a few days I had made up my mind what my course of action should be. Since those horrifying killings, a policy of stark repression had been regarded as the only solution. Police measures had touched the people at every vital point. Repression and the discontent that it raised had fed into each other. This embattled atmosphere must be lifted. That was to be my policy.

My first act was to lift the curfew. No single measure had done more harm than this prolonged interference with the normal life of the people. The dissatisfaction had permeated to all levels of society. It dislocated trade and commerce and had disrupted normal social and intellectual life. It had created a pervasive climate of fear. I believed that the device of curfew had never worked in any civilised society, except for very short periods of time to meet some special aspects of an emergency. I firmly overruled the police when they protested.

I also asked the police to cancel all the elaborate arrangements to protect my person in public places. I asked for an end to lorry-loads of armed police when I travelled by road and gave orders that there was to be no display of armed protection when I visited any public place or institution. At this point, the District Police Superintendent, an Englishman, raised the flag of revolt. He approached his superior officer, the Deputy Inspector General of Police, to prevent me from going further. The matter finally reached the Executive Council Member, Sir Khwaja Nazimuddin, who was in charge of law and order. I received information from a reliable source that Nazimuddin had pleaded with the Governor to remove me forthwith from the district as it was clear that out of sheer impulsiveness, combined with ignorance, I was bent on dismantling all the protective framework that the police had so painfully built-up over the years and that if I was allowed to continue in my reckless career he should be ábsolved of all responsibility in the matter. I also learnt that the Governor had turned down his plea, saying that he had chosen me specially for that disturbed district to bring back normalcy and I must be given a chance to use my initiative. He had also pointed out that since the most exposed person in undertaking these new initiatives would be myself, I should know what I was doing. I heard nothing officially but the resigned attitude of the Police Superintendent confirmed my information. I now felt free to go further in doing what appeared to me should be done. I had called for the removal of excessive police arrangements for my protection as I wanted to show the public that I was no longer afraid that there would be any more terrorist attacks. This, I felt, would help to establish a fresh bridge between myself and the public. I may note here that Sir Nazimuddin was a very influential person, but could not prevail against the formidable Governor. Nazimuddin, in later years, after the partition of India, became Pakistan's President (1948-51).

My next step was to remove the heavy restrictions which had been imposed on all those who were suspected of being terrorist sympathisers. These embraced some of the most prominent men of the district — senior lawyers, zemindars, intellectuals in the teaching profession. Some of them had their movements severely restricted. Some fled the district and were afraid to come back for fear of being arrested and their fears were well founded. These police measures kept the entire middle-class who formed the backbone of the leadership of the people seething with discontent. The fact that some of them were being deprived of their means of livelihood added to the misery of the situation. I asked for all these restrictions to be removed. I announced that all who wished to come back could now do so without fear of harassment. Along with the lifting of the curfew, this new measure had the immediate effect of giving back freedom to the people which was their normal right.

There was another area which called for action. I visited the educational institutions. First I visited the Midnapore College. It did not require much intelligence to note that the place was honeycombed with student spies planted by the police. The police were frank in admitting the extent of their operations. I visited the collegiate school where Peddy had met his death. I asked for these police operations to cease forthwith so as to allow students the proper atmosphere to carry on with their studies. Within a short time the change in these institutions was clearly noticeable.

At the end of my first year I heard that the Governor would be relinquishing his charge to return home and I thought it would be a good idea to invite him to see for himself the changes which had taken place in the district. He readily agreed to come for a day. I received him at

Kharagpur and took him to the bank of the Cossye river where a colourful *shamiana* (awning) had been put up. It was a bright winter day. I had asked all the district notables to come to the reception which was followed by a luncheon. There was no need for any police supervision. It was an enjoyable party, free and friendly. After he had gone I wondered what he thought about his decision to send me here or what his Executive Counsellor, Sir Nazimuddin, thought about the "folly"!

Having done these various things I took to travelling all over the district going into the most remote villages seldom visited by high officials. Any excuse was good enough for me. It might be the opening of a new library or a primary school, a prize-giving in boys' and girls' schools, some fair or other, opening a dispensary, anniversary of some noted leader, and so on. My wife was often by my side, sometimes making improvised speeches on my behalf. By the time I left the district three years later hundreds of marble plaques commemorated my visits to the various villages — this was a special weakness of the people of the district as well as presenting citations in very inflated and flowery language. In all these travels I never once met with any incident affecting my personal safety. I realized that merely removing these restrictions and giving back the conditions of normal life would not be enough to heal the deep wounds inflicted on people who had had very little to do with terrorist activities. In addition, I knew I must take up things which would show how deeply I was in sympathy with the people and make them feel that I was one of them.

One of the greatest sons of Bengal, indeed of India, came from this district in the last quarter of the nineteenth century — Iswar Chandra Vidyasagar. Though his name was a household word in Bengal, nothing had been done so far to commemorate his memory in his district of origin, either in the village in which he was born or at the district headquarters. I thought a Memorial Hall providing facilities for educational entertainment and as a social centre would be appropriate. I started raising funds. When I was sure the funds would be adequate, I invited Professor Radhakrishnan, the world-famous philosopher, who later became the President of India, to lay the foundation stone. I had known him when I was a student in the Calcutta University. His presence helped to dissipate the remaining gloom and gave credibility to the efforts I was making to restore normalcy to this afflicted district. Two years later (1939) when the construction of the hall was completed, I had the privilege of having the presence of one of the greatest men that modern India has produced, poet Rabindra Nath Tagore. He was in a declining state of health yet he came, as my wife had been one of his favourite students at his Viswabharati University. It was indeed a memorable occasion for the whole district as it was for my wife and I. He wrote to me on his return inviting us to visit him. His letter which I framed hangs on my wall. Perhaps it is getting a little faded but remains one of my precious possessions. Soon after, poet Tagore passed away. The Vidyasagar Memorial Hall stands as a monument to his memory also. I renovated the house in which Iswar Chandra was born. It has now become a place of pilgrimage for natives of Bengal.

Another incursion into the cultural field, mainly at the instance of Sajanikanta, a noted literary critic, who had also helped in arranging Tagore's visit to Midnapore, was the publication of the works of the great Bengali writer Bankim Chandra Chatterjee. He is known all over India for his song "Bandemataram" which became the National Anthem after India's independence. This publication remains the most authoritative edition of his works.

At the end of a period of two years I was asked if I wanted a transfer. The district was regarded as too difficult for anyone to manage for more than two years. I asked for another year to complete some of the things I had taken on. As the time came for me to leave I received an invitation to open a big fair. My wife received a threatening letter telling her to dissuade me from going. I did not heed the threat and she accompanied me to the gala opening. Nothing happened but some time later my wife received another letter from the same source saying that they had desisted from doing me any harm after they had seen her face!

Just before leaving, I had to arrange a meeting of the Home Member, Nazimuddin, with some 'State' prisoners (under Regulation III of 1818) who had been brought to the Hijli Detention Camp near Midnapore before release. Regulation III was usually applied to prominent revolutionaries against whom there were no charges of any violent acts but who were suspected to be the intellectual revolutionary leaders. I had a preliminary meeting with them. What struck me most was their appearance and bearing. They were all middle-aged and looked more like gentle, venerable, learned college professors than men out to destroy the mighty British Empire. They had been in detention for many years. Some of them had gone to prison when young and were now about to come out with their youth gone. For some of them the prospect of imminent release was a little bewildering — how would they pick up the threads of normal life again? It was like coming into bright sunlight after long confinement in a dark room. I always had admiration for men of this kind — men who were prepared to sacrifice all for the sake of a cause.

My first acquaintance with political prisoners detained without trial was in 1925 when in the temporary absence of the Superintendent, the District Civil Surgeon, I was placed in charge of the Midnapore Central Jail. I met a young group brought in under the newly promulgated Bengal Criminal Law Amendment Act. I used to see them on my daily rounds and talked with them. They asked for reading material. I started supplying them with whatever I could lay my hands on from my library. This came to the notice of the District Magistrate and he was not amused! Three years later when I was acting as District Magistrate of Hooghly, I met one of these political prisoners and had a long walk with him and we talked of many things. He was on his way to village internment. The Police Superintendent was courteous enough to bring this meeting to my notice as District Magistrate!

In retrospect and across a span of many years, I might say that there was a basic difference in approach between the British and Indian members of the Indian Civil Service. "The men of the Service" writes Philip Mason, a British contemporary of ours, "were chosen and trained on Plato's Principles as Guardians who would work in the light of their own vision, of the Good and the Beautiful". Reducing Plato's "Good and the Beautiful" to the realities of the modern age, this was no doubt what the British Civil Servants conceived their role to be. On the other hand, we the Indian members of the Service, looked upon ourselves as "Servants" of the people in the true sense of the word. Not that we did not often stray from the true path nor that we emulated the high example set by "The Servants of India Society"* At any rate we did make "Service" a fundamental principle in our unwritten Code of Conduct. That is perhaps why we proved a source of strength to the Government of India when independence came.

*The Society founded in 1905 by Gopal Krishna Gokhale admitting as members only a small band of highly selected and rigorously trained men, who took vows of poverty and of life-long service to the underprivileged in society.

Chapter 3

The War Years (1939-1945)

India's War Effort

I BECAME Revenue Secretary to the Government of Bengal in January 1940. Apart from various routine duties I had to keep an eye on emergency relief operations though the primary responsibility for these was with the District or Sub-divisional Officers. In Bengal such emergencies arose mainly from crop failures to famine, or sudden floods from the Ganges and the Brahmaputra, and sometimes from the small tributaries of these rivers breaking their banks. But another factor which called for attention was the havoc caused by communal riots. The Morley-Minto reforms by which seats in the Legislatures were reserved for the two communities were a reflection of the growing tension between Hindus and Moslems.

During my first year there was a serious outbreak of riots in Dacca in which many lives were lost and a lot of property damaged. I went with my Minister, Bijoy Prasad, to see for myself how the local authorities were dealing with the situation. J. R. Blair, the same man who was my District Officer at my first posting in Mymensingh, was the Commissioner. He openly resented our visit as encroaching on his authority. It has not yet dawned on some of the senior ICS officers that the 1935 Act had brought a change in which provincial autonomy came more under the aegis of the Executive to the Legislature. This reduced the authority of the Service officers. But another reason for Blair's resentment could have been personal. That a junior officer like me, though under the authority of the Minister, should be questioning him on his handling of the riots must have been a little galling!

As a result of a serious flood in the district of Tipperah, I travelled with the Chief Minister, Fazlul Huq, and the Supply Minister, H. S. Suhrawardy, to the district because the area was important for both of them politically and they were anxious to nurture the electorate. The wrangle between the two over who should give what orders was quite unbecoming.

Meanwhile, the war had been going on in Europe, the Middle East and North Africa almost imperceptibly. There was hardly a ripple to be felt in Bengal. It was towards the end of the second year, 1941, that Indian refugees from Burma came to the Assam border, most of them making for Bengal or the south of India. Soon the trickle of people became a flood of families. With the refugees came many problems. Thousands died on the journey to the border which would have brought them through inhospitable hills and jungles infested with malaria. Lack of food, water and shelter added to their hardships. And towards the end we were confronted with many cases of extreme malnutrition and other serious illnesses. Many were women and children without the male members of their family and they did not know where to turn. I met the first batch at Chandpur which was the stopping-point for the refugees coming from Assam. The condition of some of those people, particularly those suffering from starvation, was too tragic for words. I organized camps near Calcutta to give the refugees improved medical attention, and I also planned their dispersal and relocation from these camps. But many had only a few days to live.

Japan entered the war in December 1941. The Allies in the Far East had not anticipated the sudden entry of Japan into the war nor the rapidity with which Japanese forces occupied practically the whole of South-east Asia. This shook the world. But what electrified the people in India more than anything else was the naval engagement in the Pacific in which the British battle-cruiser *Repulse* and battleship *Prince of Wales* were sunk by the Japanese Kamikazes. The political and psychological repercussions in India were immediate. We felt exposed.

The military authorities were completely unprepared for any direct attack on India. Certainly in Bengal there were no visible signs of any special defence activity. Apparently it was thought that India's first defence lay in far distances. Anyway, in the East, the British were based in their "impregnable fort" at Singapore. In the West the deployment of the large Indian force in Africa and the Middle East would be a sufficient guarantee of safety. The possibility of aerial bombing had not been seriously considered.

In 1942, when the Japanese captured Singapore and their bombers appeared over Calcutta, there wasn't a single fighter to intercept them. There were two consecutive bombings before a squadron of fighter planes was brought to Calcutta. Since there was no suitable airfield and not even an airstrip for these planes, the Red Road in the Maidan (an arterial road for Calcutta traffic) had to be requisitioned. Japanese bombers were challenged and as far as I remember one bomber was brought down.

As Revenue Secretary, it was my responsibility to maintain liaison with the military and requisition whatever property they needed in Calcutta or in the countryside for their military operations.

The Japanese bombings created an emergency situation of a kind we had had no experience of before. Calcutta had a population which was the largest in India (over seven million in those days) and the most transient. It was the biggest centre in the country for manufacturing industries and for export trade in jute, tea, etc. Labour came mostly from Bihar and Orissa, two of the poorest of the neighbouring provinces. The business magnates came from upper India, apart from the Scots who traded in jute and the English in tea. Marwaris dominated the business community but retained their homes in their place of origin which was the State of Rajputana. The intellectual Bengalis considered themselves too sophisticated to put their hands to trade, commerce, speculation in shares or manufacturing industries. Many of those who did ended up in bankruptcy courts. To hold the morale of such a transient population who had no deep roots in a city under enemy attack became a problem for us of the first magnitude. It was my responsibility to devise the necessary measures to cope with the situation. As a first measure we put up thick brick walls on the ground floor of all buildings as a protection against shrapnel and we also gave instructions to put paper over glass windows following the experience in Britain. Various kinds of propaganda and publicity were undertaken to reassure the public about their physical safety. Food centres were opened under Government management to keep the food supply going in case the private markets closed down. Officials of the three railway systems — Bengal Nagpur, the East Indian and Assam Bengal — were called in to discuss measures in case there was panic and a mass exodus. Arrangements for running special trains were made. But these arrangements could only be of limited benefit — helping those who were better off and who had homes to go to elsewhere.

At this stage, I asked the Government to invest me with executive powers to handle the situation. I was appointed Director of Evacuation. I organized a chain of temporary camps for refugees, 12 kilometres apart, which was a day's walk, with all necessary sanitary facilities and a full complement of helpers for cooking, cleaning, etc. After I had set up half the number of camps needed to carry the refugees right up to the Bihar and Orissa borders, the Finance Secretary told me that he had no more money left in the budget and that I must wait till orders were received from the Government of India for the printing of unlimited currency notes. But my evacuation plans were, in the event, not needed.* The Japanese bombers stopped their incursions, either

*In fact, the chain of refugee camps, which had nearly broken the Indian Treasury, was never used.

because of the British fighters or because they had developed other strategies for conducting the war in South-east Asia. A few months later I received urgent requests from the military command to requisition a large area in the Feni sub-division in the Noakhali district where I had been District Officer years ago. The land was required for building an airstrip from which bombers and fighters could operate and fight the Burmese theatre of war. I was given a month to clear the area which was thickly populated, mostly people involved in agriculture. Removing these people at such short notice would create untold suffering. But this was war; it had to be done. I asked the District Officer to compensate these displaced people as generously as possible and clear the area within the appointed time. This was done. Millions of rupees were handed out. I asked for no accounts from the District Officer leaving it to him to answer questions when raised. At that time, obviously, the normal rules of administration had to be relaxed. The airstrip made it possible to release the Red Road for the use of the Calcutta public. The Feni airstrip was very effectively used to bomb advance Japanese positions near the Indian border.

As the war progressed, the Air Force needed more and more bases in forward areas. But these were in very sparsely inhabited locations and did not call for any requisition by the Bengal Government.

That year brought tragedies to several homes in Bengal. Our contribution to the war effort was mainly in airmen, officers and doctors in the army. Some were taken prisoner by the Japanese and several were killed in air crashes.

The Midnapore Cyclone (1942)

ONE AFTERNOON I was sitting on the verandah of my house in Calcutta. A high wind was blowing. I noticed the jagged storm clouds and remarked to my wife that a severe storm was coming. It did not reach Calcutta, so for quite a few days we were unaware that a storm had struck. When the news came through we realized it was a worse cyclone than the ones to which these areas were accustomed. But we did not know that it was a catastrophe of any magnitude until a messenger asked to see me at my house. He introduced himself as Haripuda Bhoumik a "proclaimed offender" from Contai (i.e. one who could be arrested by anyone without any warrant, being a person wanted by the police). He described what had happened in the whole of the Contai sub-division. He thought I was the only man who could save the people in this great calamity and showed great courage in coming forward, knowing I could have handed him over to the police.

I took the next train to Midnapore to see the District Magistrate and get all the information. The District Magistrate could give me only a general idea of the areas affected. He had not been able to get there since all the road communications had been disrupted. The worst affected areas were in the Contai and Tamluk sub-divisions of that district. I left by car for Contai. Soon I found the only passable road completely blocked by fallen trees. I ordered a lorry-load of men with axes and ropes to go ahead of me. It took several days to hack a way through to the area. The Sub-Divisional Officer met me and took me to his official residence. It was a solid house but with half the thick roof ripped off by the force of the storm. He described the havoc of the cyclone. The sea embankment was breached in many places and tidal waves fifteen to twenty feet high had swept across the countryside killing thousands of people and hundreds of thousands of cattle. Whole villages disappeared without a trace. The people had taken shelter on house-tops, then they moved to tree-tops and finally met their death when the water engulfed them. Rescue work was impossible until communications were restored. The seawater had polluted most of the freshwater sources. People were dying for lack of water. My visit to Tamluk was a repeat of what I had seen earlier but the area affected and the intensity of damage to life and property was less.

On my return to Midnapore, I met the District Officer again but found him most unwilling to take any special steps to deal with the calamity. I had known him before. He was the type that the British "diehards" liked best, one who had no sympathy with national sentiments or national movements. I also found that the Bengal Government, like Talleyrand's Bourbons, "had learned nothing and forgotten nothing'. The district had lately gone through much political turmoil. This officer had been specially selected to deal with it. He believed in direct, unsubtle methods. He arrested many people who were detained without trial. He felt that this natural calamity was what these people fully deserved!

To understand the situation properly it is necessary to recount the political developments in the country since the declaration of war in September 1939. In declaring war, the British Government, then under Prime Minister Winston Churchill, had also stated that India would be going to war. This they had done without any consultation with the Indian political leaders. Even the Indian Executive Counsellors of the Viceroy had not been informed in advance. Several Indian Counsellors resigned in protest. The conflict with the Congress was renewed afresh. The Congress refused to associate itself with the war effort and demanded a declaration of full independence for India. The British tried various ways to placate Indian public opinion. The Stafford Cripps Mission came in March 1942 with the proposal that India would be declared a fully independent Dominion as soon as the war ended and that the Constitution would be framed by elected representatives of British India and nominees of the Indian States. The proposal was rejected by both the Congress and the Muslim League; by the Congress because it did not comply with their demand for immediate independence and by the Muslim League because it did not meet their wishes for an independent Muslim State. On August 8 1942, Congress passed the "Quit India" resolution which was to be followed by mass civil disobedience.

Nowhere in India did the "Quit India" resolution take a firmer root than in this highly politicised district with its long disturbed history. Within a few months thousands joined in civil disobedience and sought arrest. By the time the cyclone came, many thousands were in prison. Some of the interior administrative units had become completely disorganized. The police stations were burnt down and all administrative officers had been withdrawn or had left their posts because of the insecurity. Most of these areas were hit by the cyclone. Once I had got a general idea of the nature and extent of the calamity as well as of the political situation in the district I had to decide how I should proceed in dealing with the whole problem.

I realized that the District Officer would be a hindrance to any relief operations and his authority must be superseded. I therefore decided to assume direct responsibility. I asked my Minister, in addition to my duties as Revenue Secretary, to be appointed Relief Commissioner for the cyclone-affected areas. I wrote out the order of my appointment myself and sent it to the Minister for transmittal to the Governor. This was agreed to and I assumed charge immediately.

I had first to devise ways to send relief to the areas which were out of Government control. I asked the Ramkrishna Mission, well-known for their humanitarian work, to take charge of the supplies I provided. The Mission readily agreed to distribute these. I also accepted an offer of help from the Friends Society of Britain (with Horace Alexander as their leader) whose representatives were already on the scene after hearing of the calamity. The Friends Society group set out for the disaster area in a boat loaded with supplies. I remember this because one of them was a young woman and I wondered how she would be able to stand the hardships in those extraordinary circumstances.

My second task was to try and repair the breaches in the coastal embankment through which the salt water still flowed in. It was a tremendous undertaking. The Embankment ran for nearly 60 or 70 miles and the breaches had to be repaired between the high tides. I gave the two sub-divisional officers of Contai and Tamluk a blank cheque to somehow accomplish the task. Several thousand men and women drawn from the countryside had to be engaged for those few days that were left to us.

Many years later, after India's independence, when I was Ambassador in a far-away country, the news was brought to me that the Public Accounts Committee of the Indian Parliament had raised the question of accountability for these expenditures and of the expenditures for the Feni Airstrip in 1942 also (though Feni was no longer within Indian borders). However, wisdom must have prevailed for I heard nothing more about accountability for expenditure in those eventful days.

Next, I turned my attention to the political leaders who had sought arrest by civil disobedience and who would have been the natural leaders in organizing relief in those difficult circumstances. I asked that those arrested without trial for civil disobedience be allowed to come to my aid in organizing relief. This was immediately opposed by the District Officer. The Chief Secretary, the same J. R. Blair of Mymensingh and Dacca experience, was wholly on his side. The matter came to a head. A meeting was called at the Secretariat at policy level which did not include the District Officer but the Chief Secretary was there to speak for him. It was presided over by the senior Minister, Shyamaprasad Mookherjee, and it was a stormy meeting. I stated the case that it would be rational to release the political leaders so that they could take responsibility for organizing proper relief in their own areas. The Chief Secretary thought that in the prevailing political situation this could create an uncontrollable law and order situation. Shyamaprasad Mookherjee, a great patriot and a powerful leader, overruled the Chief Secretary. He took the matter up with the Governor but to no avail. These decisions were put into effect much to the chagrin of my friend, the District Magistrate, and the relief operations shifted into top gear with the help of the local leaders.

Yet, when one comes to think of it, how far do these relief operations, undertaken nationally or internationally, in this or any other similar case, go beyond tiding the stricken people over the worst of the crisis. The business of rebuilding their lives from nothing remains with them as a daunting task for long years. It is indeed surprising how tenacious people are and how strong is their will to survive. This reminds me of something I have never forgotten from those days. As I was slowly making my way along a road of fallen trees, I noticed a sapling growing out of one of them. There it was, waving and dancing in the sun, a beautiful symbol of man's desire and capacity for survival!

The Bengal Famine (1942-43)

AS IF THE CUP of misery for the people of Bengal was not full with what had already happened, now came one of the worst famines in the history of this famine-haunted country. The total failure of a ripening rice crop in Bengal was the first signal of the coming disaster. Recurrence of famines had been a feature of Indian life because of the dependence of the main crops on the unpredictable and variable monsoon rains. As far back as 1867, the duty of the State to provide famine relief had been recognised and an elaborate famine relief code was adopted to prevent famine and also to ameliorate famine conditions through public works. Since the famines were generally localised, supplies were brought in from neighbouring areas.

This time, there were some very special features due to the conditions created by the war. Burma was the main supplier of rice for India and Burma was cut off because it was occupied by Japanese forces. The military demands on transport for the Burmese theatre of war influenced the movement of supplies from the surplus areas. But there was also a general failure on the part of the administration to deal with the situation.

The Bengal Famine Commission Report (1945), presided over by a distinguished member of the ICS, Sir John Woodhead, a former Governor of Bengal and a man of great integrity and competence, brings out the facts in their stark reality. "It had been for us a sad task", the report says, "to enquire into the course and the causes of the Bengal famine. We have been haunted by a deep sense of tragedy. A million and a half of the poor of Bengal fell victim to circumstances for which they themselves were not responsible. Society, together with its organs, failed to protect its weaker members. Indeed, there was a moral and social breakdown, as well as administrative breakdown." The Commission then refers to the lack of a "margin of safety" either as regards health or food resources of a considerable section of the Bengal population. They cite the more immediate causes that I have mentioned — sudden loss of Burma rice and the military needs of transport and food. But the Report goes on to lay a large portion of the blame with the Government of India and the Government of Bengal. "But after considering all the circumstances", says the report, "we cannot avoid the conclusion that it lay in the power of the Government of Bengal, by bold, resolute and well-conceived measures at the right time, to have largely prevented the tragedy of the famine as it actually took place . . . (Also) the Government of India failed to recognise, at a sufficiently early date, the need for a system of planned movement of food grains . . . The Government of India must share with the Bengal Government responsibility for the decision to de-control in March 1943 . . . The subsequent proposal of the Government of India to introduce free trade throughout the greater part of India was quite unjustified and should not have been put forward. Its application, successfully resisted by many of the Provinces and State might have led to serious catastrophe in various parts of India." The Commission then refers to the apathy and mismanagement of the governmental apparatus, both at the centre and in the province, and goes on to say, "the public in Bengal, or at least certain sections of it, have also their share of blame. We have referred to the atmosphere of fear and greed which in the absence of control was one of the causes of the rapid rise in the price level. Enormous profits were made out of the calamity, and in the circumstances profits for some meant death for others. A large part of the community lived in plenty while others starved, and there was much indifference. Corruption was widespread throughout the province and in many classes of society. The total profit made in this traffic of starvation and death was estimated by the Commission at 1500 million rupees. Thus, if there were a million and a half deaths by famine, each death was balanced by a thousand rupees of excess profit!"

Though I personally was an important administrative figure in the provincial government, I have to admit that, by and large, the observations and judgment of the Famine Commission were objective and correct. As I recall those tragic days I often wonder what more I could have done and did not do. Since I was responsible for emergency relief why did I allow things to get so far out of hand without protest? Why did I not cry out louder when the *aman* crop failed and ask for planned supplies from other provinces? The only reason I can now think of is that the Midnapore cyclone had taken place at about the same time as the *aman* crop failure and I had immediately plunged myself into the relief operations in that district and perhaps this had fully occupied my mind. Maybe I was also under the impression that since crop failures were a recurrent phenomenon in that part of the country, this one would also be taken care of in the normal way under the Famine Code procedures. No one in the Bengal administration, including myself, could then conceive that things would take the course they did. India had regularly imported all its rice needs from Burma with the average import of rice being three and a half million tons a year. Following the fall of Burma, the immediate military demands on available food and transportation on the Burma front were not foreseen. The Japanese advance had overshadowed everything else. When the decision to declare a famine was taken, other priorities were allowed to divert attention. I recall H. S. Suhrawardy, then Minister in charge of supplies in Bengal, once remarking to me that the Central Government Food Secretary, Brigadier General Wood, had passed orders for seven train-loads of food grain from the Punjab, but that the trains never materialised and never came was not his business or the business of any other officer at the Centre! Whatever measures the Central Government took to ameliorate the situation had the effect of worsening it. The de-control of prices was conceived to encourage private traders to move food grain from surplus to deficit areas. But this immediately pushed up costs and the grain could only reach those who had the means to pay high prices. This led to profiteering. Fear and lack of confidence in the ability of Government to manage the situation became pervasive and in areas like Calcutta those who could afford to do so went in for hoarding on a large scale. The proposal of free trade made by the Central Government, also with the same objective of ensuring a freer flow of the surplus to Bengal, had the effect of drying up market supplies in the surplus provinces as the authorities there wanted to ensure the safety of their own people before those in Bengal. There appeared to be a pall of doom all around, while the social life in Calcutta went on as before as if nothing was happening.

Eventually, starving people began to arrive in Calcutta in hundreds of thousands on foot or by the suburban trains. They came to die *en masse* on the railway platforms and on the Calcutta streets. The dead bodies lay rotting in the sun and the municipal authorities could not cope with such numbers of corpses. The Famine Commission estimated the deaths at one and a half million; the department of Anthropology of the Calcutta University at three million, four hundred thousand. Whichever estimate was correct it was a holocaust. Those of us who were in any way involved cannot escape a profound and abiding sense of guilt.

Chapter 4

India's Director-General of Food

September 1943-46

IN SEPTEMBER 1943 when the force of the Bengal famine was gradually abating, I was called to Delhi to take over a newly created post as Director-General of Food for all India covering the entire sub-continent including presentday Pakistan and Bangladesh. The Bengal famine had clearly brought out the mistakes and failures of the Central and Provincial administrations, and these mistakes and failures had to be urgently attended to.

It will be relevant to set out the nature of the Government of India at that time. The 1935 Act was still the framework for the make-up of the Government. The Congress-affiliated Executive Counsellors resigned in 1939 because of the declaration of war involving India. The Indian Executive Council Members of the Viceroy were now persons of eminence but isolated from the political life of the country — landlords, businessmen, practising lawyers. The Member in Charge of Food was Sir Jwalaprasad Srivastava, a businessman from the United Provinces (now Uttar Pradesh). He was a happy-go-lucky man, rather fond of pleasures of the flesh but also very shrewd and intelligent. He knew his political limitations. He was the *bête noire* of the Congress and under constant attack.

The rapid Japanese advance in South-east Asia and the fall of Burma had a profound effect on Indian thinking. Though the Japanese advance was halted on the Assam border, the prospect of an imminent invasion hung in the air. Gandhi called upon the British to leave Indians to deal with the Japanese by non-violent means. The "Quit India" movement started at this time. It was also the time of the Congress resolution demanding immediate independence, the launching of the mass Civil Disobedience Movement, the arrest of Gandhi and the members of the Congress Working Committee and the emergence of Jayaprakash Narain at the head of a short-lived guerilla movement. It was a time of struggle and it went on for nearly two years — till the end of the war. But at the war's end, there was a period of relative political calm.

This was the situation in which I was to organize my new duties. In October of the same year, Lord Wavell took over from Lord Linlithgow. I had met Lord Linlithgow only once and that was in Calcutta when I was sent to explain to him the situation about the Burmese refugees. He had the reputation of being a proconsul of the old type. At the formal banquet which was *de rigueur* even in those difficult war times the ceremony was still based on the lines of the British royalty. When the time came for the ladies to retire after the banquet every one of them had to make an elaborate bow before the King's representative, the Viceroy! I found Lord Wavell to be an entirely different type. He dispensed with these royal trappings and was always approachable. The article in the Encyclopaedia Britannica describes him as "a distinguished soldier who found great difficulty in communicating with any warmth to all but his closest associates." I suspect this article was written by one who had been near Wavell in India and had been denied that warmth. As time passed, and especially from about the end of 1944, my work brought me in contact with him from time to time. He was interested in problems of food and famine. Two occasions I remember in particular. Once at the beginning of 1945, a problem arose when seventeen ships with military supplies for the Burma war zone were waiting at Bombay harbour to be unloaded. It so happened that three ships with food abroad were also waiting to be unloaded. General Lindsay was the Military Officer in Charge of Supplies for the army in Burma while I had to feed the civilian population. Whose ships should have priority in unloading? We could not agree because we both thought our respective needs were more urgent. The final court of appeal was the Viceroy. We sought an audience with him and as we entered he looked at both of us and then asked me to explain. I said that unless the food ships were immediately unloaded, Bombay would be without food within two days and this would also affect port labour. He took up his red and blue pencil and immediately scrawled the orders "food ships first"; he did not ask General Lindsay what he had to say. On another occasion, in 1945, after the Japanese forces had withdrawn from the Dutch East Indies, Indian forces had been sent to clear up the situation in Java. We had arranged to send a shipload of food and other necessary supplies for the civil population. When the ship was nearing its destination I received a telegram saying that it had been boarded by a party of Royal Dutch sailors who regarded the supplies as contraband. I asked to see the Viceroy and as a result he asked his Private Secretary to send an immediate message to the War Office in England to have the ship released.

The Bengal Famine Report came out in 1945 and what the Report brought out about the inadequacies and inefficiencies of the Central and Provincial administrations was already clear enough to us. The purpose in creating this new post of Director-General of Food was to try and correct these mistakes and see that such a catastrophe did not overwhelm us ever again. The de-control of food grain prices and the proposal for free trade in food grain, which had had quite the contrary effect on the movement of supplies to the deficit areas, had to be immediately cancelled. The introduction of all-India rationing was suggested and we called on the advice of experts who had made such a success of the rationing system in Britain. The chief expert who came was one by the name of French. I went to meet him on his arrival at Karachi. We were both the guests of the Governor, Sir Francis Mudie, a formidable old member of the ICS whose word was law within his kingdom. I was then trying to locate food surpluses in the main producing centres of which Sind was one. At the discussions, I mentioned this and the scale of surplus I wanted. I was firmly told that he and nobody else would be the deciding authority. As we broke up French confessed to me his dismay at this attitude. On my return to Delhi I brought this to the notice of the Viceroy. A meeting of the Governors was called. The Viceroy left Sir Francis in no doubt that the Food Department was the authority which would decide such matters. There was no further trouble getting the surplus from Sind.

Lack of planned movement which had been a major factor in the Bengal tragedy had to be replaced by a much more efficient strategy. All this had to be attended to and special officers had to be appointed — a railway movement expert, a storage expert, a rationing expert and so on. To provide for coming events (i.e. to establish a machinery which would be in a position to give early warning) the whole administrative system at the Centre as well as in the Provinces had to be geared up. We had already a special representative for each of the five or six regions into which the Central food administration had been organized. This was the structure when I arrived. But I also found that some of these Regional representatives were not effective. I had to replace some of them with men who were more competent and had the background needed for the job. To establish better and closer liaison with the officers of the Provincial governments, I had to travel extensively. Air communications were then still very inadequate and mix-ups were a regular feature.

Apart from my executive duties as Director-General of Food I was nominated a member of the Council of State at the beginning of 1944. The Council of State was the Upper House of the Central bicameral legislature and every member had "Hon'ble" prefixed to his name. India being a Federation under the 1935 Act, the Council had appointed members from the States as well as elected members from British India. In addition, some seats were reserved for nomination by the Governor-General. I was one of his nominees representing the Food Department which was then very much in the spotlight as a consequence of the Bengal famine. One of my tasks as a member was to explain the lessons we had learnt from the famine and the fresh measures we were adopting to prevent the recurrence of such famines. The most colourful figure in the Council of State was India's Commander-in-Chief, Sir Claud Auchinlek. He was resplendent in his military uniform. He had won acclaim for having halted Rommel in North Africa when Wavell had failed and he had saved Egypt. He was replaced (September 1942) by General Montgomery having asked Churchill for a brief respite from the counter-offensives against Rommel to acclimatise his soldiers and give them further training. Paradoxically, when Montgomery asked for an even longer period of training before committing troops against Rommel, Churchill agreed. All this became public knowledge and only increased Auchinlek's prestige in India. Among the good speakers was Hriday Nath Kunzru. He was one of those members of the Servants of India Society who have left behind a rare record of dedicated public service in India's modern history.

The war was not coming to an end as was the saga of Netaji Subhas Bose. He had formed his "Provincial Government of Free India" (1943-45) and his Indian National Army had marched up to the borders of India. In 1945 he died in an air crash at Taipeh as he was flying to Tokyo. His companion, Habibur Rahman, survived to tell the story. As the Germans surrendered in May and the Japanese in September the political situation in India again became disturbed. Sir Jwalaprasad, who was the Member in Charge of Food, found it increasingly difficult to face the Lower House of which he was a member. He resigned and asked the Food Secretary, an Englishman, to take over his portfolio. This man was a good speaker but apt to be somewhat patronising in his ways. He was shouted down at his very first appearance. It was then that I was nominated to take his place to represent the Food Department.

The Lower House was very different to what I had been used to. The elected members were mostly Congress or Muslim League representatives. There was much cut and thrust between themselves, and between both of those parties and the government members.

The first interesting event I remember was the election of Speaker. The incumbent was Sir Abdur Rahim, a veteran Muslim League leader who was standing for re-election. The new candidate put up by the Congress was Mr. Mavlankar. Abdur Rahim even though he was strongly supported by the Government was defeated by one vote.

Towards the end of 1945 I was very involved in my responsibilities as Director-General of Food as there were signs of an impending crop failure in southern India. As soon as I brought this to the notice of the Viceroy he asked me to accompany him in his plane to Bangalore from where he could monitor the developing crisis. We were there for two days and we went to see the parched countryside and the great water reservoirs for irrigation constructed a century ago, all dried-up now. When we returned to Delhi he asked me if I needed any special help. I asked for the setting up of a Committee of Secretaries of all the Departments involved which I would preside over and which would coordinate all aspects of relief. This was a novel initiative based on the Bengal famine experience. It was greatly resented by the Secretaries who were all senior in rank to me. However the crisis did not materialise and a famine was averted.

About this time I had engaged a well-known businessman to buy all available surplus rice on our behalf from Nepal, which was our main source from abroad after the fall of Burma. Purchases worth several crores of rupees (10 million = 1 crore) were made and brought to a railhead connecting Nepal with India. But before any of the bags could be loaded the monsoon rains came and in the space of a few days the mountain of bags rotted and had to be written off. Questions were raised in the Lower House about the loss and I had to explain. My statement of the circumstances in which the loss had occurred was accepted.

With the end of the war, the Labour Party came into power in Britain with Attlee as Prime Minister. The sequence of events in India — the fresh elections to the Central and Provincial legislatures to test the demand for a separate Muslim State, the proposals of the 1946 Pethick-Lawrence Mission and their rejection, the "Great Calcutta Killings" (Hindu-Muslim riots in which about 7,000 persons died), Wavell's Interim Government with Indian leaders followed by a wave of communal riots in East Bengal — this was the background against which we all had to work. With the new elections the Lower House assumed a more dynamic character. Bhulabhai Desai was the Congress spokesman in the House but he wasn't an impressive figure as he spoke like a lawyer in Court. Nawab Liaquat Ali Khan, the Muslim League spokesman (who had been a fellow-student with me at Oxford) did not claim any brilliancy in the debates but, with his common sense, became a dominant figure in the House (he later became Prime Minister of Pakistan). One of the finest and most incisive debaters was Dewan Chamanlal.

My contact with the new Congress Members left a painful memory. With the situation in south India still precarious, our Food Department came up with a proposal to send a delegation of prominent political leaders to Washington to ask for assistance in grain. As I had to present the leaders with that proposal and win their approval, I first went to see Mr. Jinnah who was then a member of the Lower House. He had a reputation of inflexibility. As soon as I explained the proposal he said that the Muslim League would participate. I then went to see the Chief Whip of the Congress Party (I would rather not name him) because the leader, Sarat Chandra Bose, was too ill to receive me. He was totally opposed. At the very mention that we should ask for a grant from the USA, he became angry and said on no account should we go begging even if thousands of people were to die! The proposal got no further.

In making that proposal we had been encouraged by the visit of ex-President Herbert Hoover who came as President Roosevelt's Personal Representative to survey India's food needs at the end of the war. I had shown him round for nearly a fortnight and he visited different parts of India and saw things for himself.

When the Interim Government was announced the news got around that one of the first acts of the Congress leaders would be to de-control food grain prices. At that time of scarcity this would have been most unwise. I was asked by Sir Jwalaprasad to see the Congress leaders to dissuade them from this course. I first saw Jawaharlal Nehru and explained to him at length the history of the case — what had happened during the Bengal famine and why we had re-imposed the control. He listened patiently for nearly an hour and then asked me to see Govinda Vallabh Panth as he said he really did not understand the problem. I immediately took a plane to Lucknow. I had not known Panth before. He was a giant of a man both physically and in leadership qualities. After listening to me he said lightly "you people think you know best!" But he remembered. In later years whenever I visited India as FAO's Director-General he would receive me and discuss the food problems affecting not only India but the world as a whole. He was a regular reader of my "Monthly Letter" to Ministers of Agriculture and often raised various points for discussion.

At the end of the year (1946) I was selected by the Interim Government presided over by Jawaharlal Nehru to open the Indian Embassy at Washington. Before I left I went to pay my farewell call on Lord Wavell. In spite of his preoccupation with the political developments he had encouraged me to keep in touch with him. He said he would write to his friend Field Marshall Morgan, Head of the Joint Allied Military Mission in Washington, to introduce me. His personal kindness to me all through those years in Administration had meant a great deal. I admired his qualities. He had a wonderful memory. At any formal function when he had to deliver a speech, it would be printed and distributed beforehand but those reading his paper would be aware that the spoken and written word matched exactly yet he never looked down at a word! In the preface to his anthology of English poetry *Other Men's Flowers*, which was published after he left India, he wrote that he had included only those pieces which he could recite from memory — the anthology covered the entire range of English poetry. If he had not been a General of the British Army, he would have found a place as Professor of Poetry at Oxford University!

The Great Divide

THE YEARS 1946-47 were a most critical period in India's history. The differences between the Congress and the Muslim League came to a head during these years. Jinnah discarded the somewhat ambivalent attitude of previous years and came out strongly in favour of a separate State for the Muslim majority provinces. In June 1945, Wavell tried to have the two Parties come together to form a Government but the negotiations broke down on how the Ministers were to be chosen. When the Labour Party came into power at the end of the year and openly committed themselves to independence for India, the belief still persisted that Jinnah was using his demand for a separate Muslim State in order to extract more concessions from the Congress. This was certainly Wavell's belief and Gandhi and Nehru also held the same view. The elections of 1946 were intended to test Jinnah's theory. But the results were inconclusive. The frustration of the Muslim League manifested itself in the great Calcutta communal riots. The two Parties were left more divided than ever.

Following the Calcutta riots, Wavell as a last resort set up in Interim Government in September that year which with certain changes finally emerged with seven Congress members and five Muslim League members in the Cabinet. The portfolios had to be so apportioned as to give both Parties sufficient strength to act as a restraint on the other. Foreign Affairs and Home Affairs went to the Congress, and the critical portfolio of Finance to the Muslim League. This delicate balance had the effect of accentuating rather than lessening the differences. There were often deadlocks on important policy issues.

The Muslim League, however, was not much interested in foreign policy matters at that stage and Nehru, with his unrivalled knowledge of foreign affairs, had a freer hand in initiating new foreign policies for India. It was against this background that Nehru decided, as a major policy initiative, to open an Embassy in Washington which was, after all, the most important capital in the world. He was anxious to have a means through which the Congress' attitude on national and international questions could be made known to the world. The sympathetic attitude of President Roosevelt towards India's political aspirations (which he had made manifest in his dealings with Churchill over India) was also a compelling reason. Until then, India's important Diplomatic Offices abroad — the High Commissioner in London and the Agents General in the USA and China — though headed by Indians, were regarded by the Indian political leaders as no more than appendages of the British Establishment. Sir Girija Shankar Bajpai, Agent General to the USA, was one of the most brilliant members of the ICS. He was a fine debater, an excellent draftsman of constitutional and other similar documents requiring a good command of the English language, and had great juristic skill. He was reportedly very much in demand in the drafting of the Charter of the United Nations at the San Francisco Conference which he attended as a delegate of India. He also attended the Conference at Quebec which adopted the Charter for FAO. He was a valued member of the Far Eastern Commission which was set up to monitor the administration of Japan by the US Supreme Commander, General McArthur. When I took his place in the Commission, the French representative in the Commission who had worked with Bajpai spoke to me of his outstanding qualities. But the Indian political leaders looked upon him as a *protegé* of the British and therefore not to be wholly trusted as an authentic voice of the new India.

It was at this point in time that I was asked to proceed to Washington to take over from Bajpai and to open an Embassy to which full diplomatic status was to be accorded. The name of the new Ambassador was also made known. He was Asaf Ali, a trusted lieutenant of Nehru, who was already a member of the Cabinet of the Interim Government. The idea of sending me ahead was to establish contact with the State Department and clear all necessary formalities. I was also to get the Embassy Chancery adjusted to function in accordance with the status of an Embassy.

A few days before I was due to leave for Washington I went to call on Asaf Ali. I had met him before in connection with the proposed All-Party delegation to Washington, to negotiate the deal on the supply of food grain. We did not take to each other on that occasion and now I found him somewhat cool. I felt he did not like the idea of my being his Minister Plenipotentiary. I learnt that he had the impression that I was being sent to keep an eye on him. This certainly could not have been Nehru's idea and it was quite a shock to me. I never asked for this assignment and I had never conceived diplomacy as being my field. I was as inexperienced as Asaf Ali himself. I had been assured before this appointment came that I was soon to take over as Secretary of the Food Department which was the highest position I could attain in my line. I understood that he made representations to Nehru to let him choose his own man but failed to have the order changed.

Asaf Ali's attitude towards me was not entirely personal. It was also an expression of the general attitude of Congressmen towards the ICS. What Nehru said in his *Discovery of India* explained that attitude best: "and not the Viceroy only but the British members of his Council, the Governors, and even the smaller fry who function as Secretaries of Departments or Magistrates. They speak from a noble and unattainable height, secure not only in the conviction that what they say and do is right, but that it will have to be accepted as right, whatever lesser mortals may imagine, for theirs is the power and glory. . . ."

It was only to be expected that when there was going to be a total transfer of power, from British to Indian hands, one of the main props of the ancient regime — the ICS — should be regarded with suspicion and hostility. The transfer of power was done by a constitutional process not by revolution or violence. India was a big country with a population bigger than that of Europe or of North and South America together. The diversity in languages and customs of the people in different parts of the country needed an administrative machinery which was already well-established and of a high standard of efficiency. The ICS provided such a machinery even with its Indian element only. No one understood this better than Nehru and his right-hand man Vallabh Bhai Patel. In leading the new India, Nehru provided the statesmanship and the vision, and Vallabh Bhai supplied the realism and strength that was needed. Vallabh Bhai insisted that a trained administrative Service like the ICS was essential, especially during the transitional stage. He believed that the Indian element of the ICS would be sympathetic to the ideals and aspirations of the people and should serve the country well. His own steady hand and the iron will of Vallabh Bhai were to be Nehru's greatest assets in dealing with the most difficult problems that had to be faced. He made the fullest use of the Indian members of the ICS in various old and new fields of work from then on. Bajpai, though withdrawn from Washington, became his first Secretary-General of the Foreign Office, and he was allowed to recognize the entire Foreign Service to serve the new independent India. He brought in KPS Menon as Foreign Secretary, despite the fact that he had been India's former Agent General to the Chiang Kai-shek regime (1943-46). Nehru appreciated his exceptional intellectual abilities and sent him as Secretary-General of the delegation to the UN General Assembly in 1946 which was led by his sister, Vijay Lakshmi Pandit, to raise the subject of apartheid in South Africa. Though I was "a small fry" he did not hesitate to overrule Asaf Ali when he objected to my going as his Minister to Washington. Since then the Indian members of the ICS have had the fairest deal in all respects and have been fully used in the service of their country. In 1979 — after 32 years — the last of the 416 Indian members of the old ICS retired. Some have distinguished themselves and only a few have been found wanting in loyalty or dedication. We leave behind a proud record.

Chapter 5

An Interlude in Diplomacy

In Washington as a Diplomat

I LEFT for Washington in January 1947, after waiting for several days in Bombay for the TWA plane which was making its inaugural flight. We stopped over in Cairo, Athens and Rome. We crossed the Atlantic by the shortest route, Shannon-Gander, and landed at Boston for entry-check into the USA. The sight of New York in the clear evening festooned with lights opened up the promise of a new world. We travelled on to Washington and arrived at midnight in a blinding snowstorm.

I did not meet Bajpai, India's Agent General to the USA, on this occasion. I therefore found myself totally exposed to the swirling political and diplomatic problems that then invested the capital of the world. As I look back, I think my only qualification then was a certain mental discipline I had acquired over the years as an administrator in my country. In the ICS, I had had opportunities in different capacities, calling for legal, administrative and even journalistic proficiency. I do not recall having much difficulty in adjusting myself to the new surroundings in Washington. I have never been a party-goer but soon, I had to get used to the incessant round of cocktail parties prevalent in diplomatic circles. The late Senator Greene, Chairman of the Senate's Foreign Relations Committee, was a veteran party-goer. Someone once asked him which party he was going to next that particular evening. He replied "if you tell me at which party I am now, I will tell you where!" This about represented my situation. I have been a teetotaller all my life;

I used to carry one drink in my hand to last through a party to show my active participation in the merry-go-round.

I found the office premises Bajpai had established quite adequate for a Chancery. The residence he had purchased was one of the best in Washington and was the envy of other Ambassadors. There had been a separate Supply Mission for purchases found necessary during the war and the man in charge considered himself to be under direct orders from Delhi. The Consulates-General in New York and San Francisco also worked on their own. One of my tasks was to bring these offices in line with the policies of the new Embassy. Since our relations with the British Embassy had been so close and I was a neophyte in diplomacy, I readily accepted the invitation of the British Embassy to join with their officers in a weekly meeting to survey the general political developments and to discuss any specific economic or political problems in which we might all be interested. I did this in full awareness that I represented my country on the verge of independence with its own outlook on certain international problems. There was no conflict in my mind nor was there any attempt on the other side to influence my judgment in any matter.

Since I left India some very important developments had been taking place. On 20 February 1947, Prime Minister Attlee announced in Parliament that it was the intention of the British Government to transfer power into "responsible Indian hands by a date not later than June 1948" and because there was no assurance that agreement could be reached on the form of government for the new India, he planned to transfer power to some of the provincial governments if necessary. He further announced that Wavell would be succeeded by Lord Louis Mountbatten as Viceroy.

Soon after, Ernest Bevin, the British Foreign Secretary, came on an official visit to Washington. I wanted to ask Bevin about this announcement and in particular why Wavell was being replaced. Even as late as December there had been no talk of any such possibility. Bevin arrived with a toothache and this was front-page news in the Washington papers! I thought my best approach would be to commiserate with him on his toothache. But, from his rather surly response, I realised that I had a long way to travel before I became a diplomat. I obtained the information I wanted from his aide, John Strachy, who told me that since power was to be transferred by a particular date and seeing that Wavell was against partition, it was thought desirable to have someone in his place who had an open mind on the matter. Indeed, it was quite well-known that Wavell had all along thought that Jinnah would finally come round with some more concessions from the Congress side. In this he was in agreement with Gandhi and Nehru. Gandhi had gone so far as to offer, much against the will of the other Congress leaders, control to the Muslim League provided they agreed to a united India. The terms of Attlee's announcement did not in any way differ from what Wavell had been asking for. The question of partition was not mentioned; the words were left deliberately vague. A definite date by which the transfer was to take place was also what Wavell had recommended. Wavell was a man of great integrity. I wonder if his true worth was ever recognised.

Mountbatten's arrival and the swift developments that followed are well-known. On August 15, the transfer of power was announced. Freedom came at midnight. Nehru's speech on "the tryst with destiny" reverberated throughout the world. But before the reverberations had died down, death and destruction seized the land. The mass movements in the east and in the west got under way. Killings and massacres on both sides put the "Great Calcutta Killings" of the previous year into the shade. I was frequently asked by the press in Washington and New York to explain what was happening. Our joy in having independence at last was mixed with deep sorrow at such happenings.

Our Ambassador Asaf Ali arrived. He was very conscious of his high position. He was the first native Ambassador of soon-to-be independent India. He liked ceremony and more than that he loved to be adulated by those working round him. This did not increase his popularity. In one of his first morning meetings with the senior staff members, he asked for a world map to be spread before him on the wall. A standard Mercator projection map, showing the world with Europe in the centre, India on the right side and the Americas on the left, was brought in. He was not satisfied with this. He wanted India to be shown in the middle and since no such map had been published the world map had to be removed!

The first two months of the Ambassador's time, after his presentation of credentials to President Truman, had to be spent in making the usual formal calls on his numerous colleagues and receiving their return calls. Day-to-day contacts with the State Department were left to me while he reserved for himself interviews with the Secretary of State. My relations with the State Department were very cordial. The attitude of the State Department reflected the general attitude of the country's interest in India. I noticed a feature in the State Department administraion which was different from what I was used to — the tendency of the lower level officers to dispose of matters coming from the various Embassies without consultation with superior officers. On the other side of the coin, the instructions of superior officers often did not seem to percolate down to the lower levels.

At that time, India's prestige abroad was at its highest. Gandhi and Nehru were household names in the United States. The mighty industrial and business communities were straining at the leash waiting for markets to open in India. Enquiries were pouring in from banks, steel, locomotive, computer and other industries. There was an air of expectancy. This was just the kind of atmosphere that suited the style of our new Ambassador.

Another matter which naturally absorbed his attention was the Arab-Israeli war which was being urgently discussed at the United Nations. Being a trained barrister and backed by our natural sympathies for the

Arabs, he soon came out as the champion spokesman for the Arab cause. The entrance hall of the UN building, then at Lake Success, was filled with photographs of our Ambassador with the Arab leaders. He brought to bear upon the Arab cause all his legal skills and he received high acclaim as a result.

The Indonesian Debate

APART FROM the Arab-Israeli war, another question came up before the General Assembly that interested our country. Before the Japanese forces surrendered to the Allies in September 1945, the Japanese Government had installed Soekarno, one of the Indonesian leaders as the head of a government in Java and Sumatra. When the Dutch returned to their old colony after the cessation of hostilities they would not agree to this *fait accompli* and started police action to dislodge his government. They arrested several Indonesian leaders and took other action to re-establish their authority. The matter came up before the Security Council. Soekarno asked India to represent the Indonesian case as Indonesia was not a UN member, being still Dutch territory juridically. My Ambassador and I were attending a cocktail party when a telegram came to ask the Ambassador to represent Indonesia's case. The person who had been briefed in India on the question had been unavoidably delayed in London. The Ambassador turned round to me to ask me to take over. Apparently he thought that it was *infra dig* for him to take someone's place at short notice in such an important matter. Since the brief had also remained with the person detained in London, we had no official instructions from Delhi to act on. It was evening when I rushed to the Chancery to see if there was any material which I could use. There was nothing except some telegrams which had already appeared in the newspapers.

The next morning I took the plane to New York for the United Nations Security Council meeting on the Indonesian question. I felt like an innocent abroad! I was delayed for an hour at the airport where I saw Ambassador van Kleffens who was to represent the Dutch cause in the United Nations debate. He was walking up and down in a pensive mood, which rather heartened me since he did not seem so sure of himself either. At the Indian permanent delegation to the UN, we had a very bright young man, Samar Sen, who was Secretary to our delegation. He put together the relevant material that was available in the UN Secretariat on the Indonesian case and this information gave me a good idea of the issues involved. I then felt somewhat more confident about representing the Indonesian cause.

The Security Council debate went on for several days, ending inconclusively. Though the case had first been presented to the Council by Australia, the debate became very much an exchange between myself and van Kleffens for the Dutch. This man was a noted expert in international law. (His successor in 1949, when the Indonesian debate was again taken up, was van Rooyen, who was also a good debater. But both men were aware that world public opinion was against them.) No conclusions were reached during this debate and the whole question had to be postponed for nearly a year for the situation to clarify. But the exposure of the Dutch police action in Indonesia did much to shape the future. This was my first experience of the United Nations and the Security Council.

Later, during my FAO days, I came in close contact with the Dutch Government, especially through my friend Sicco Mansholt, Minister of Agriculture. I found them to be most co-operative and positive in their approach to the problems of the developing countries. In recent years, they have taken the lead in assistance programmes and have reached, and even exceeded, the 0.7 per cent GNP target for development assistance laid down by the United Nations General Assembly. But I always felt that the UN debate on Indonesia in 1947 left some scars at the highest political levels in that country.

My personal relations with the Ambassador continued to deteriorate. Our mutual antipathies were too deep to be papered over and Delhi soon became aware of it. It became a question of withdrawal for one or the other. A senior officer of the Embassy who had been to Delhi on leave came back to say that I was to stay. But soon, other news seemed to contradict this. I decided to go back and sent a telegram to Delhi asking for other duties in India. In reply I got a message from the Minister of Agriculture, Dr. Rajendra Prasad (one of our foremost leaders who later became President of the Constituent Assembly and India's President), to come and take over as Secretary to the Ministry. I had worked for him for a short time before leaving for Washington. I accepted.

This diplomatic interlude in Washington was an interesting one from many aspects and I did not regret having come.

Chapter 6

Return to India

The Ministry of Agriculture

AFTER A STAY of less than a year in Washington, I returned to Delhi in December. I returned to a divided and truncated country. It had been hoped that some viable formula would be found to satisfy the Muslim League and still retain the unity of India. But finally partition had to be accepted as the only acceptable solution. Only Gandhi remained unreconciled and decided when his formula was rejected to withdraw from Congress politics. Vallabh Bhai had played a crucial role in convincing Nehru and other Congress leaders of the realities of the situation.

Though I had been one of India's spokesmen abroad, when I came face to face with the stark realities it was another matter. The Radcliffe Award, which laid down the demarcation lines in the Punjab and Bengal and which was submitted only two days before the declaration of independence, had caused millions from either side to leave their homes and professions and go in search of security. The trek of eight to ten million refugees from Pakistan to India and from India to Pakistan with mass killings, starvation, sickness and exposure was a tragedy of incalculable dimension. This was specially true in the east as the Hindus from east Bengal, unlike those from the Punjab or Sindh, were mostly destitute. They poured into Calcutta and the suburban districts and made the railway platforms and the streets of Calcutta their temporary homes. The authorities were swamped under and with the uncertainties of the times they were unable to mount measures to receive and look after this large influx of refugees. My own forefathers had lived in east Bengal for many generations. Many people we knew were among the refugees. They were mostly middle-class professional families. This tragic situation remained unresolved for a long time. Even today, after all these years, the stream of refugees has not altogether dried up.

Apart from these tragic mass movements the problem of accession of the Princely States and more specially the events in Kashmir over accession was uppermost in the minds of the political leaders. The incongruity of the patchwork-quilt of Princely States, big and small, so laboriously woven by the British in their own interest and which juridically were to revert to their original status after the termination of the British paramountcy, was obvious. The Princely States, realizing that the upsurge of the masses would soon sweep them away, faced up to the inevitable and came to a *modus vivendi* with the Congress Government. Vallabh Bhai supported by Mountbatten acted swiftly. His Secretary, V. P. Menon, was instrumental in carrying out this policy. Within a very short time, over five hundred States were merged into greater India. The only two where problems arose were Kashmir and Hyderabad. By the terms of the partition the States could opt either for India or for Pakistan or remain independent. After some hesitation the ruler of Kashmir opted for India on 26 October 1947. This led to local unrest among a section of his Muslim subjects who were encouraged by Pakistan since they considered Kashmir to be legitimately their territory under the principles of the partition. The Pathan tribesmen from Pakistan joined in this local unrest and such was the situation at the end of the year.

When I arrived back in Delhi from the United States I stayed with General J. N. Choudhury, Chief of Staff of our Defence Forces. Though I was not directly involved in what was going on, it did not require a Sherlock Holmes to guess that the Kashmir question was causing a lot of concern to the Cabinet and the General was often called away to emergency meetings at short notice.

I began to pick up the threads of my new assignment as Secretary to the Ministry of Agriculture of the Indian Union. I considered it great good fortune to have Dr. Rajendra Prasad as my Minister. I had worked with him for a while before I left for Washington. He was a man of great culture and erudition with a warm personality and free of prejudice. But my good fortune was too short-lived. Dr. Rajendra Prasad was called upon to be President of the All India Congress Working Committee, a position of crucial importance at the time.

He was to be succeeded by Jairamdas Daulatram at the Ministry of Agriculture. I had never met Jairamdas Daulatram before. He was a disciple of Gandhi, editor of his newspaper *Harijan* and had been interned for civil disobedience. These were no doubt very good credentials for promotion to a Minister but I discovered that these were not necessarily the qualities which made a good Minister. His date of arrival was announced and on the day, he was expected to arrive by train some time after midnight. Normally, I would have gone to welcome him. But looking at the hour of arrival, I thought that this was no occasion for turning out an Honour Guard. I sent my most senior officer in the Ministry to receive him and take him to his residence for the night. I was soon made aware of my mistake and my failure to meet him was interpreted as an expression of ICS arrogance. Also, the "reputation" I had acquired in Washington as a difficult person to work with had followed me. The Asaf Ali affair had reached the ears of Gandhi and this might have influenced my Minister.

The Ministry of Agriculture was lodged in the South Block of the Secretariat in the spacious rooms intended for an Imperial Secretariat. Immediately after assuming office, the Minister conveyed to me his desire to address the staff; he specially mentioned that the *Chaprasis* (uniformed messengers inherited from the British days) should not be left out. Since there was no hall to accommodate large numbers, arrangements were made in an adjacent quadrangle below. A big *satranchi* (jute matting) was spread out with a few chairs. He asked for the chairs to be removed and we all squatted on the floor. Simple living and high thinking! He addressed the assembly for nearly forty-five minutes. The burden of his speech was that India had achieved freedom but work still remained to be done to weed out the ICS which represented the old regime. For forty minutes he pointed to me sitting by his side as the representative of the evil which would have to be exorcised. I was his Principal Officer! This harangue was certainly not conducive to discipline or good administration, apart from its bad taste.

It was not long before he addressed himself to raising the standard of administration. Among other things, he asked that all files dealing with *Chaprasis* — their leave applications, their grievances — must be sent to him personally for disposal. I told him such matters were normally disposed of by the junior officers and this would be a waste of his time. But being a true democrat he declared that everybody would have the same attention from him. It took only a few days for his office to fill with files which he couldn't find time to attend to. What the poor *Chaprasis* thought about this attention from the Minister, which in reality only complicated life for them, I can only guess.

Another matter which brought our relationship to a breaking point was inter-secretarial meetings which were meetings between Secretaries of the different Ministries to discuss matters of mutual interest. He asked me not to attend these meetings without first clearing with him what I was going to discuss and what views I was going to express. I explained to him that these meetings were not intended to circumvent the Ministers but only to clear up points which would take inordinate time if dealt with in memo form. I assured him that policy issues or important administrative matters were always brought to the Ministers' attention. He was not satisfied. I then explained the status and responsibilities of a

Secretary *vis à vis* the Minister. I pointed out that a Secretary drawing the salary that I did must be expected to justify it by exercising some responsibility. I insisted that I must have full freedom to express my views and exercise my judgment and it was then the prerogative of the Minister to accept or reject my views. Nehru had some pertinent remarks in his *Discovery of India* about the attitude of the Congress leaders towards the Service when they assumed power in the Provinces under the 1935 Act:

> "... Yet since we had some opportunity, however limited and restricted, in these Provincial Governments, we wanted to take advantage of it in the fullest sense. But it was a heart-breaking job for our Ministers who were overwhelmed with work and responsibility, and could not even share these with the permanent services, because of the lack of harmony and the absence of a common outlook ... The Ministers were supposed to set an example in plain living and economy in public expenditure. Their salaries were small, and we had the curious example of a Minister's Secretary or some other subordinate belonging to the Indian Civil Service drawing a salary and allowances which was four or five times the Minister's salary. We could not touch the emoluments of the Civil Service. Also the Ministers would travel second-class by railway train, or even third, while a subordinate of his might be travelling first or in a lordly saloon in the same train."

This was the situation in the 1930s. Since then there had been closer association and understanding between the Congress leaders and the Chief Executives of the ICS. Vallabh Bhai had restored a proper focus to this whole problem when India became independent. The gap between the Ministers and the Service had also been narrowed. Ministers no longer travelled second-class on trains and, if their salaries still lagged behind those of the Service, the various extras more than made up the difference. But my Minister still lived in the 1930s and his attitude had not changed with the times.

Towards the middle of the year there was an FAO/ECAFE (UN Economic Commission for Asia and the Far East) Regional Conference at Bangalore which I attended. Our Finance Minister, John Mathai, was the Chairman. The Conference, following a resolution of the UN Economic and Social Council (ECOSOC), set up a Working Party to study rehabilitation of agriculture in the war-devastated countries in east and south Asia. I was named as Chairman of that Working Party. I readily accepted and my Minister readily agreed though we both had different reasons. I agreed because this would give me an opportunity to travel and work and also I would be getting away from the tense atmosphere of the Ministry for a while. The Minister agreed because my absence would give him the freedom to review his relationship with me.

Chapter 7

On the Road

China and Formosa

TOWARDS the end of August 1948, I arrived in Shanghai to take up my new assignment. It was a big undertaking to be accomplished within a very short time. The countries to be studied were Burma, Ceylon, China, India, Pakistan, Indonesia, Indo-China, Malaya, Philippines and Siam (I am using the names of the countries as they were then known). We were to examine the question of rehabilitating and raising the agricultural production of these countries which had been affected by the war. This was to be done through improved supplies of fertilisers, transport vehicles, machinery and spare parts, equipment for power generation and irrigation, pesticides, etc. We were to indicate priorities in requirements and to study what opportunities existed for local production of the necessary requisites. I headed the Working Party assisted by a noted Chinese industrial engineer, C. Y. Lin, who was my Deputy. We were to be joined by a group of experts from FAO and ECAFE and specialists drawn from different countries. But as the regions to be covered were so vast and air communications between them still so difficult, we divided the responsibility for the study of various countries' needs among all of us. I concentrated on China, Formosa and Japan. Japan was not yet a member of the United Nations as it was still under the authority of the Allied Supreme Commander, General McArthur. I included Japan for our study as I considered it a potential source of supply for the agricultural requisites that would be needed.

Shanghai was the seat of ECAFE and I made this city my headquarters. I stayed at the Cathay Hotel on the Whampoo harbour, property of the Sassoon family. The first realization of the effects of runaway inflation came when I had to pay my dinner bill at the hotel. It came to over three million Chinese yuan (normally it should have been 15 yuan to the dollar). The level of inflation was unbelievable reducing the whole social structure almost to rubble. At the hotel I met a group of American businessmen. One of them who became a lifelong friend took me out to the countryside to see the living conditions of the rural people. As we neared a village, we met a group of children who ran alongside our car begging as they ran. My friend took out a heap of Chinese currency notes from his briefcase for them. But they did not want these scraps of paper and weren't even interested. The village itself presented a picture of unbelievable squalor and poverty. The people appeared to be barely able to keep themselves above starvation. Indeed, the whole country was a scene of disorder, poverty and social breakdown. Ever since the beginning of this century, China had gone through a succession of revolutions and rivalries of warlords who had divided out the country among themselves. On top of all this the Japanese invasion created havoc. The condition of the people was a result of this continued turmoil. At that moment Chiang Kai-shek was locked in battle with the Communists and Mao Tse-tung was emerging as their leader.

Shanghai itself epitomised all the ills of the country. Millions of people from the countryside left their hearth and home and came to the city seeking shelter, safety and livelihood. As a result, side by side with dancing, feasting and gay life among the well-to-do (foreign enclaves were still intact), the poor starved and were uncared for. The main streets swarmed with beggars which reminded me of my own home city, Calcutta. The beggars here were most methodical in their profession — the footpaths were divided among the beggar families by chalk marks. One night as I was returning from some formal Chinese reception, I saw a family stretched out on one of those chalk-marked squares keeping an all-night vigil. A baby was kneeling with the head touching the hard pavement in a posture of begging. It was past midnight and it was a soul-rending experience.

I travelled widely in China by train, by plane, and by motor vehicle. After organizing the work I went to Nanking which was then Chiang Kai-shek's capital, as I wished to pay a courtesy call on the Prime Minister. The door was opened by a small-sized man. Taking him to be the doorman I brushed past him and looked towards the other end of the long room for the Prime Minister. But it was the Prime Minister who had opened the door! We talked about the general political situation in the country and I explained my mission. He described the course of the civil war and invited me to come with him to the Map Room. The war maps were operated by push button with varicoloured lights and it was an impressive display. One could see the progress of the war in minutest detail. Even bridges of critical importance which had been destroyed were indicated in coloured lights. I saw this technological legacy from the American Army put to good use in other countries years later such as in Thailand and South Korea when they displayed development plans and projects.

In Nanking, I came across an Agricultural Research Institute still functioning with a dedicated staff of high-level specialists.

I was staying with India's Ambassador to China, K. M. Panikkar, one of the finest brains in our Foreign Service. He was a distinguished writer and historian and I learnt a good deal from him about the political and economic conditions in China. I attended one of his lectures at the Nanking University though I did not think that the students took in much of what he said about Indo-Chinese relations. They were politely inattentive.

Panikkar was with me when I flew to Peking for my official visit. He was to do his diplomatic round at the same time. Our plane circled round Peking Airport for quite a while and we were told that Chiang Kai-shek had just arrived and everybody would be delayed. When we landed the whole place bristled with soldiers. Chiang Kai-shek had come to meet and talk to the University authorities — there were five universities in Peking at that time — as more and more students became sympathetic to the communist cause. But he came too late. The Communist forces had already taken over the Great Wall and were at the gates of Peking. The "Forbidden City" in Peking, the walled area exclusively reserved for royalty in the past, was now given over to students who had come, along with the faculty members, from the Harbin University in the north, ostensibly to avoid falling into the hands of the Communist forces but in reality to avoid enrolment on the Communist side. After their "Long March" the students were no more pro-Chiang than before; they appeared to be only performing a ritual common in those days. The hardships of their "Long March" had embittered them and if anything they were in a rebellious mood.

I was invited to lunch by the President of the Chinghwa University, an American, John Leighton Stuart. He had been General Marshall's Advisor during the abortive discussions in 1945 between the Communists and Chiang Kai-shek. He was later appointed American Ambassador to China (1946-48). As we were going to his residence within the precincts of the University, I noticed a series of wall posters and asked my host what they were about. He said with a wry smile, "posters vilifying me as an enemy to their country". I became aware of the Chinese way of conducting a media campaign — the wall poster!

Panikkar and I stayed in the Wagon-Lits Hotel during our week's stay in Peking. The hotel was full of White Russians who had fled the Bolsheviks and were still patiently waiting to find permanent homes in other countries.

I visited all the tourist sights in and around Peking except the Great

Wall. The Summer Palace of the Manchus outside Peking was still thronged with Chinese tourists. The half-submerged marble boat with exquisite carvings, fashioned at the bidding of the Dowager Empress for her pleasure trips on the lake, still remained as a testimony to the follies of a decadent dynasty which was about to be extinguished.

On my return from Peking, my friend and colleague, C. Y. Lin, introduced me to Chinese families and I experienced warm, unbounded hospitality. Thirty-six courses for dinner were still the tradition! We visited the Buddhist temples in Hankow famous for their fascinating architecture and went to Soochow which was known for its beautiful women.

I got down to hard work in Shanghai and with an American friend visited a warehouse near the city in which a vast amount of American Army surplus stores were kept. I wanted to know how the surplus — tractors, trucks, machinery of all kinds which could be put to use for agricultural purposes — was, in fact, being used. I was told that nothing was being released for fear it might fall into enemy hands! My American friend addressed an open letter to Chiang Kai-shek which was published widely in the local papers. I still have a copy. I got the impression that whatever we might write in our report for the rehabilitation of agriculture in China would not have the slightest effect on the situation as it was developing at that time.

I made a short trip to Formosa. After a tiring air journey, I went for a hot bath to refresh myself. As I was lying relaxed in the bath, a young lady came in to give me a massage. My face must have registered a startled and outraged expression because she retreated very hurriedly!

My visit to Formosa was not very productive. The political turmoil in mainland China had already spilled over to this island. The Japanese, after their surrender, had withdrawn completely from this former colony of theirs. Their armed forces, as well as the civilian population who had made Formosa their permanent home for several generations, had all departed. But with the civil war in China rising in crescendo and with the setback of Chiang Kai-shek's forces, mainland Chinese began to pour in with the tacit acquiescence of the Americans. Chiang Kai-shek himself was poised to transfer his headquarters to this island to fight another day.

The island was in a state of utter disorder. Many Formosans came to see me to ask my help in raising the question in the United Nations of how they were to get their country back. In fact, at that time they were preparing to send a delegation to the UN General Assembly to plead their cause. But the tide was too strong against them. Their feeble voice was drowned in the cacophony of the migrating hordes. There was little room for me to carry out my mission. I noticed that the Japanese had left behind quite a high-level agricultural base. The first sight which greeted a visitor was the neat lines of terraced hillsides for agricultural operations. It was the industries which appeared to have been attacked by the American bombers. Every worthwhile industry or workshop had been reduced to a mass of twisted steel and rubble. Complete rebuilding, rather than repairs, was what was needed to bring the essential industries back into production.

Japan

I LANDED at Haneda — later to be the giant cosmopolitan Narita airport. It was little more than a landing strip then, the surroundings bearing the scars of war and the hangars a mass of twisted steel. There was an air of doom and destruction. The people appeared sulky and uncommunicative which was in contrast to the warm smiling faces of the Chinese I had just left behind.

The countries which had fought in the Pacific war including India were each represented at that time in Tokyo by a Diplomatic Mission of a special character since Japan was not then regarded as an independent country. I was received by our Diplomatic Representative who was a member of the ICS I had previously known. I went to stay with him and was able to use my short time in Japan to the best advantage. He had been quartered in the house of a nobleman, who was now living in the original quarters for the domestics. I noticed some of the old feudalistic practices still remained. Whenever the master went out or came in the domestics would line up in front of the house to bow low in unison to show their respect.

The war had ended in September 1945 and General McArthur, the Supreme Commander, was responsible for administration of the country with all its post-war problems. He was already proving himself a great Pro-Consul. He drew up a new Constitution to divest Japan of its militaristic (Samurai) traditions. The Japanese army was to be reconstituted and equipped only for maintaining internal security and defence. He retained the institution of the monarchy which he regarded as central to Japanese life and resisted all attempts to arraign Emperor Hirohito as a war criminal. He carried out a comprehensive land reform giving the right of ownership to the tillers of the soil and in this way destroying the feudalistic structure of the society which had been the foundation of the Samurai. The educational system was overhauled. It was indeed fortunate that there was a man of such Caesarian proportions, in prestige and authority, to guide Japan through this traumatic period. He stood over Japan like a wise giant. He had one weakness, also of giant proportions — his vanity — which ultimately brought about his downfall!

One of my first engagements was to make a courtesy call on the Supreme Commander. As he was not available his deputy in charge of diplomatic affairs received me very kindly and put me in touch with the

Director responsible for agricultural matters. Sitting beside the Director was a Japanese who was introduced to me as the Director's counterpart. I realized that the Japanese gentleman was a Minister without authority because the administration was under the Supreme Commander. He invited me to dinner which I readily accepted. The Director was also asked to join us but he appeared somewhat dubious. He finally accepted. He later explained to me that, under the Supreme Commander's orders, Americans, whether in a civilian or military capacity, were not to fraternise with the Japanese! As I learned afterwards this amounted almost to some form of apartheid though not on the basis of colour. Travelling by train we had to have a compartment separate from the Japanese!

One of the Japanese officers of the Department came to take me to the dinner. He was in charge of public relations. It was he who gave me all the details about non-fraternisation which he said was galling to Japanese pride. When I arrived at what I thought to be the residence of my Japanese host, a beautifully dressed Japanese lady came to the door and asked to take off my shoes. I agreed but thought it strange that such should be the custom for showing respect to guests. When I went in there were many guests seated on the matted floor with low tables in front of them and one Japanese lady-companion attending to each guest. The dinner started after several toasts and the lady by my side filled my cup every time a toast (gambai) was called; she helped me to pick up the delicacies from the table and in other ways showed her single-minded devotion to my comforts. It was half-way through the dinner that it dawned on me that this was not the residence of the Minister but a Geisha House. I found myself completely unlettered in these matters. After the formal dinner, the party was given over to western-style dancing — a concession no doubt to the American Occupation. Since I was the chief guest I had the most beautiful geisha attending to me. I asked her what her name was. She said her professional name was "the Flower of Shanghai". She had come home with the soldiers. I must say I enjoyed the party and I bade my companion farewell and went back to my quarters.

I did not let my preoccupation with work interfere with my interest in photography. I carried my camera everywhere and this was a land for photography. With my friend, the Diplomatic Representative, I visited many of the temples, shrines, parks, and flower shows. I was photographed with him by the side of the giant Kamakura Buddha. I was entertained at other Geisha Houses by my Japanese friends, but never in their homes. That was not the Japanese custom, they said, because they lived poorly and simply. I noticed the reserve of the Japanese in their daily life but I found when they went to a Geisha House for entertainment, they were very relaxed and full of fun. Their normal reserve was really a mask, a creation of generations of enforced social discipline; at heart they could be as gay as any other people.

I travelled widely and observed everything. I made as thorough a study of Japanese agriculture as I possibly could. I prepared a detailed note which was incorporated in the Final Report of the Working Party submitted to the United Nations. In my note, I observed that Japan was not only the most highly industrialised country in Asia and the Far East but also the most advanced in agricultural production. Japan's average yield for cereal crops was the highest in the world, more than double the yield in the USA. This was due to controlled irrigation and the use of high quality seeds and fertilisers, organic and inorganic. But the most striking feature of Japanese agriculture was its agricultural research: a large number of stations with independent responsibilities dealt with localised problems which arose from the wide diversity of climatic conditions. Another remarkable feature was the Extension service. Village cooperatives worked in close liaison with the Experimental Stations and agricultural technicians, one for each village or group of villages, strengthened the link by helping farmers to make use of research and kept the Experimental Stations informed of the farmers' problems. In spite of war damage, Japan's industrial capacity for supplying the needs of the Far Eastern countries in agricultural machinery and implements was considerable. In relation to my mission this was important. The Japanese agricultural practices continue to be an object lesson for all.

On our return journey from Tokyo I met my friend C. Y. Lin who was carrying a very large parcel. Lin was a calligraphist of national fame. He had just bought back some precious Chinese scrolls which the Japanese, during the Occupation in the thirties, had removed to Tokyo.

I had been only a few days in Shanghai when I received a telegram from my Minister in New Delhi asking me to proceed to Washington to be India's delegate to the FAO Conference. I left with a sense that our mission in Japan had failed. The region was still too disturbed for any systematic action or any fresh initiatives. There were hardly any national plans at that time to which our recommendations could be properly related. The countries in their written replies to our questionnaire had come forward with nothing more than shopping lists of agricultural requisites.

As I said goodbye to my friend Lin he presented me with a scroll decorated in his own exquisite calligraphy. I often wonder what happened to him and his family after the change of the regime in China.

Washington

IT WAS A LONG flight to Washington — Shanghai-Tokyo-Wake-Honolulu-San Francisco-Chicago-Washington. I was given no brief from India on the FAO Conference. To suddenly switch the mind from the crowded experiences of the past three months to something entirely new was itself a strain and added to the physical strain of the long journey.

I had been in Washington only a year before in a quite different capacity. This was not my first acquaintance with FAO. When I was Director-General of Food in India I had followed the course of events leading up to the Hot Springs Conference and then the drawing up of the Charter of FAO at Quebec. I had met John Boyd-Orr, FAO's first Director-General, on my previous visit to Washington; he was about to resign over the rejection of his initiative for a World Food Board that would use agricultural surpluses for food-deficient countries. I had visited him in his office in the modest red-brick building on Massachussets Avenue. Who could then imagine that FAO, with five or six separate buildings each more than a dozen times the size of this one would in the future be silhouetted against a Roman skyline alongside the Colisseum and St. Peter's? He then had a budget of only $5 million a year and no more than a few dozen officers. But even with these modest resources he had made a mark. The studies (1947-48) for the agricultural rehabilitation of Poland and Greece would remain a standard reference for years to come. I also knew Frank McDougall who, after persuading President Roosevelt to accept the idea of the creation of an International Organization like FAO to deal with post-war food problems in the world, sent that famous telegram to his friends in England — "We have today lit a candle, that by God's grace, will never be extinguished." He was an active staff member.

But the 1948 Conference was my first real introduction to FAO — to the work of the Organization as well as to some of its most notable figures — Viscount Bruce, ex-Prime Minister of Australia, who had been the spokesman on food and agriculture before the League of Nations, a man of unrivalled knowledge and experience; and Professor André Meyer (College de France, Paris), another notable pioneer.

As the Conference began I was elected Chairman of Commission III to deal with the technical activities of FAO — Nutrition, Agriculture, Fisheries, Forestry, Rural Welfare, Economics and Statistics, Trade in Commodities, etc. One of the Vice-Chairmen of this Commission was Professor Josue De Castro (Brazil) who later succeeded Lord Bruce as Chairman of the FAO Council. Michel Cepède (France), whose long connection with FAO was also climaxed by his Chairmanship of the FAO Council, was then a fresh young man with great verbal talent and full of erudition. He was Rapporteur to the Committee on Rural Welfare under my Commission.

I learnt much from my work on this Commission and I think I was also able to bring in some fresh ideas. We laid down general principles which should guide the Director-General in future planning. We pointed out the need to evolve projects on a four to five-year basis. We observed that the vital test of the budget of each Technical Division should not be a predetermined share of the total available funds but the priority of projects. For major projects, there should be "combined operations" with the closest possible integration of divisional and regional activities. We emphasised that extension and educational services should be properly adapted to reach the ordinary farmer. I remember making one proposal entirely on my own experience which if consistently adhered to might have given a somewhat different image to FAO. My proposal was that, while FAO should maintain a permanent core staff to ensure continuity, Member Governments should undertake to grant leave of absence of up to five years to their officials so that they could accept temporary service with FAO. On the one hand, this would bring fresh ideas and experience to FAO and, on the other, when these officials went back they could take with them international experience of value to their country. This inflow and outflow would be mutually beneficial and FAO would no longer be regarded as an ivory tower inhabited by overpaid bureaucrats.

The star figure in the debates was the Irish delegate, Charles Dillon. His witty speeches always drew crowds. The polished French which the delegate from Haiti delivered was acclaimed to be better than the French spoken in the salons of Paris.

The Conference ended in November. As I was resting from my labours a telegram arrived from my Minister in New Delhi stating that I should investigate the possibility of purchasing 5,000 agricultural tractors, investigate the prices and other details. I thought it rather strange that we should get involved in such a large venture so suddenly. However, I took the plane to Chicago which was the centre for agricultural machinery. I telegraphed back all the information I could collect and returned to Washington. I had been away from home for nearly four months and I was anxious to get back when I was the recipient of another telegram to say that I should wait for further work in the USA. My Minister's anxiety to use my valuable experience seemed to me strange especially as I knew that up to the time I left he seemed to go out of his way to denigrate my abilities in every matter. I did not hesitate any longer but took the next available plane back to Delhi.

Homecoming

I ARRIVED BACK in New Delhi by the middle of December. I rang up my Minister to tell him of my return. He asked me not to go to the office but to see him at his residence the next day. It sounded rather odd that he should ask me not to go to the office. Next day when I arrived at his residence I found him preöccupied. He did not ask me about my experiences during my four months' absence abroad nor about the FAO Conference which should have been a matter of special interest to him. The question of the 5,000 tractors was not raised. However, he soon came out with what had been his preoccupation. I was to be placed on special duty to act as liaison between a World Bank Mission from whom a large loan had been requested and the Chief Ministers of the States. In his view I was the only one who could talk in a forthright fashion with the Chief Ministers. I was duly flattered but unconvinced. I suspected that he was trying to find some way of preventing my return as permanent Secretary at the Ministry of Agriculture. When I remembered my relations with him before I left this did not surprise me. But what was surprising was that he should use such underhand means. If only he had been a little more frank before I left and told me how he felt about me I would have cooperated as I would have been happy enough to have a change. The whole affair created quite a sensation when it became known because such a thing had never happened before.

I took over my new assignment as Officer on Special Duty. I do not recollect that there was any significant call on my services either from the World Bank Mission or from the Chief Ministers. I did not expect there would be. But things started moving in other ways.

I had a phone-call from the Minister of Supplies, my old friend since the Midnapore days, Dr. Shyamaprasad Mookherjee. The Chief Minister of West Bengal, Dr. B. C. Roy had called on him and they were both coming to see me. Dr. B. C. Roy had begun his life as a physician. He had returned from England with the highest medical qualifications and within a short time became one of the foremost physicians in India. It was in one of his early years that I had his attention as a doctor when I had fallen ill as a student. We had known each other ever since in different capacities. In the early twenties, he entered politics and had come forward as a successful Congress candidate to the Bengal Legislature under the 1919 Act. After the partition of India he had taken over as Chief Minister of West Bengal. He was now on a visit to New Delhi for consultations.

Dr. Roy asked me if I would like to come as Chief Secretary to the West Bengal Government under him. The post of Chief Secretary was the highest and most responsible position for the ICS in a State. In those days the powers and responsibilities were still undiluted by politics. The offer was tempting but I hesitated for two reasons. One was Dr. Roy himself. He was a man of tremendous physical and mental resources, the most dynamic of the Chief Ministers but he stood like a giant banyan tree under which no grass could grow. I had great respect for him but I felt that one of my kind could not get along with him for too long. I asked Dr. Roy for time saying that I had just returned from abroad and I needed to consult with my wife. The other reason was the chaotic political condition in West Bengal after the partition and the situation created by the influx of the millions of refugees — the suffering and neglect and the inaction of both the Central and State governments. I felt that I would be no more than a spectator in such a situation, unable to do anything. As soon as Dr. Roy left I asked Shyamaprasad for his advice as I had great confidence in his judgment. He confirmed my apprehensions. I turned down Dr. Roy's offer. He was not pleased and I think that our friendship lost its warmth after that. An offer then came from Shyamaprasad himself that I should take over as Secretary of his Ministry. I pointed out to him that this would be a repetition of the same kind of thing that Jairamdas Daulatram had done which we all deplored. The incumbent was an old colleague of mine, we had entered the ICS together and he was known to be a very able officer. While these discussions were going on I received a message that an additional post of Secretary was contemplated in our Foreign Office and it was suggested I might be interested. My old friend, K. P. S. Menon, was Foreign Secretary and it was inconceivable that I should in any way be responsible for weakening his position, even remotely. I declined.

I continued as Special Officer to the World Bank Mission and then in late January 1949, I was asked by our Foreign Office to be ready to proceed to New York immediately to lead the Indian delegation to the Economic and Social Council of the United Nations. I was to take the place of Sir Ramaswamy Mudaliar who for some reason was unable to go. I was happy to have this break but enquired what my next posting would be after I had finished the ECOSOC session in the middle of March. I was told that I would be going back to India's Foreign Service and would be posted as Minister to the Indian Embassy in Washington. I would be *Chargé* till the arrival of the new Indian Ambassador, Madame Vijay Lakshmi Pandit.

I accepted the offer after discussing it with my wife. Our youngest child had been ill for some time and the household affairs had to be wound up. She readily agreed to take these upon herself and let me proceed at almost a week's notice. She joined me with the family in Washington after a month. Thus I came to the end of one pathway in my life. I was now to embark on an altogether new career with new problems and possibilities.

I have written in some detail about the relationship with my Minister. I am not interested in vindicating myself or belittling him. But I do wish to place on record the kind of differences that arose between a senior member of the ICS and a Minister under the new dispensation. The new All-India Service — the IAS — since set up is still undergoing a process of adjustment. Now there is a trend for the States to discount the value of the people in the All-India services because they are searching for greater autonomy for themselves. If my experiences help towards providing some useful guidance in finding a true harmony between politicians and Civil Servants, I shall feel more than justified in having related them.

Chapter 8

The Cold War

The Chill begins

IN 1947 when I was in Washington I felt the first icy winds of the Cold War but without a trained ear did not quite understand the significance of this first sign. On March 12 1947 President Truman, in his address to the Joint Session of the Congress enunciated what is now known as "The Truman Doctrine" and as *Chargé* of the newly opened Indian Embassy I was among the Diplomatic Representatives seated in a specially reserved enclosure on the Congress floor. In addition to Greece and Turkey, he promised unlimited aid to "other nations threatened by communists" anywhere in the world and I took this to be more a flight of fancy than a policy seriously intended. When Secretary of State, George C. Marshall, in his address to Harvard University (June 5, 1947) advanced the idea of a programme of European economic self-sufficiency supported by US assistance, I simply regarded it as an act of unprecedented generosity arising from the traditional humanitarianism of the American people such as had also been seen at the end of World War I (Hoover Mission). The implications of these offers, however, became clearer when I returned to the scene this time after a lapse of a year.

If there was any real understanding between Roosevelt and Stalin on how the post-war world should be built, that understanding was not in evidence after Roosevelt's death on April 12, 1945. Russia's rapid consolidation of influence over the countries of eastern Europe first raised the danger signal for the West. But this could be explained away as a process of consolidation after the upheavals caused by the war. However, when signs became visible of the same process of advance in southern and eastern Europe — Turkey, Greece — the West could no longer remain inactive.

The Civil War in Greece (1946-49) began over the question of the restoration of the monarchy but the communist guerillas also came into action with the open support of the neighbouring communist countries like Yugoslavia, Bulgaria and Albania. Turkey had turned to the West when it felt threatened by Russia's territorial demands on its Caucasian frontier and for the freer use of the Dardenelles for Russian warships. Seen against this background the Truman Doctrine appeared to have meaning, at least as far as Greece and Turkey were concerned. In asking for funds from the Congress, Truman said that the funds were needed to aid Greece and Turkey in their resistance to "attempted subjugation by armed minorities and by outside pressures". The Greek Civil War continued till the autumn of 1949. As a result of the Tito-Cominform split the Yugoslav border was closed and the Greek communist guerillas, lacking outside support, could no longer hold out in their mountain strongholds and finally surrendered.

Despite the unquestionable generosity with which the USA poured in assistance to stimulate the economy of Europe — about 3 to 4 per cent of its GNP, running to many thousand millions — it was clear that the inspiration was part of a strategy to stem the advance of communism in western Europe. An economically prostrated Europe, it was argued, would be an easy prey to advancing Soviet influence.

I mention these two outstanding events of 1947 because of their repercussions on the debates in the UN General Assembly which I attended later in the year as a member of the Indian delegation.

I was still *Chargé* of the Indian Embassy when I was invited as a member of the Diplomatic Corps to witness the ceremony of signing the NATO Treaty on April 4 1949 in Washington. This brought the Cold War out into the open. Dean Acheson who had succeeded George C. Marshall was the prime architect. It was an unprecedented step for the USA to become involved in a peace-time military alliance. Article V of the Treaty provided "... an armed attack against one or more of them in Europe shall be considered an attack against them all ..." To a general observer it was an occasion for sadness that so soon after two of the most devastating wars in history preparations had to be made for another.

The Economic and Social Council session was interesting in many ways. On most matters I had to prepare my own brief. This gave me a sense of freedom which I specially valued. We covered a wide variety of subjects, from World Economic Development and Human Rights to Forced Labour, Translation of the Classics, Equal Rights for Men and Women for Equal Work and Effects of Chewing Cola Leaf. But the most important debate I remember was on economic development of the underdeveloped countries. The American delegate, Willard Thorpe, opened the debate with a reference to President Truman's Point IV. In his Inaugural Address under paragraph (4) Truman had said:

> "We must embark on a bold new programme for making the benefits of our scientific advances and industrial progress available for the improvement and growth of underdeveloped areas. More than half the people of the world are living in conditions approaching misery. Their poverty is a handicap and a threat both to them and to more prosperous areas. For the first time in history, humanity possesses the knowledge and skill to relieve the suffering of these people. The USA is pre-eminent among the nations in the development of industrial and scientific techniques. The material resources which we can afford to use for the assistance of other peoples are limited. But our imponderable resources in technical knowledge are constantly growing and are inexhaustible ... in cooperation with other nations we should foster capital investment in areas needing development. Our aim should be to help the free peoples of the world, through their own efforts. We invite other countries to pool their technological resources to this undertaking ... This should be a cooperative exercise in which all nations work together through the UN and its agencies wherever possible ... With the cooperation of business, private capital, agriculture, and labour in this country, this programme can greatly increase industrial activity in other nations and raise substantially their standards of living."

One or two points in this statement raised questions in my mind. Coming so soon after the great Marshall Plan I wanted to know why was the offer of assistance to the underdeveloped countries so halting, so general, so limited. Also, what was the significance of the words "our aim should be to help the free peoples of the world?" Was this again an echo of the Cold War? "In cooperation with other peoples", meaning the same peoples who were then being aided by the Marshall Plan, these words seemed to take away some of the seriousness from the new bold programme.

In response to the American delegate I described the Point IV programme as "neither bold, nor new, not even a programme", following Voltaire's quip on the Holy Roman Empire. I confess that in saying this I was somewhat undiplomatic, somewhat inclined to levity. After all, the USA had been very generous in giving help to other countries. My remarks caused great dismay and indignation among the Americans but my statement was greeted warmly by Charles Malik (Lebanon) and Boris (France) when they met me after the meeting. The next morning we began a debate on "Survey of Forced Labour and Measures for its Abolition". This was a Report by the American Federation of Labour which made special reference to conditions in the USSR. In my statement I said that while forced labour should be condemned wherever it existed there were other skeletons in the cupboards of other countries which should also be recognised. (I was making a veiled reference to segregation in the USA.) At that point a telegram was placed in front of me by my Aide. The telegram was from New Delhi chiding me for my remarks on Truman's Point IV. However, New Delhi did not pursue the matter further and I was thankful.

I took up my position as Minister on the termination of the ECOSOC session at the end of February and I awaited the arrival of Madame Pandit from Moscow where she was then Ambassador. She was coming with the prestige of being Nehru's sister and also with a high reputation for her performance, two years before, as India's spokesperson in the UN General Assembly on apartheid in South Africa. I have already mentioned the NATO ceremony I attended before she arrived. It was also about that time that the Commonwealth Ambassadors went to the airport to greet Winston Churchill who was paying a courtesy visit. As I extended my hand to him he clearly showed his disapproval of India by talking to the next person while taking my hand.

Madame Pandit was appointed leader of the Indian delegation to the UN General Assembly again that year. I was included as one of the delegation. It was a momentous session. All the significant developments which followed the Cold War found full expression. The effect of the Civil War in Greece was debated. The Greek representative was one of the most eloquent speakers. Alex Bebler (Yugoslavia) had much to say about happenings in Yugoslavia including the split with USSR. Turkey pleaded its case with subdued eloquence because of the difficulty of language. Vyshynsky and Gromyko (USSR) held the centre of the stage. In the midst of all this turmoil came the news that the USSR had tested its first atomic weapon. This, for a time, overshadowed the entire scene. The babel was hushed. People spoke in whispers. The entire world balance had been changed. The memories of Hiroshima and Nagasaki were still hanging in the air. Where would this nuclear race end?

The Korean Debate

ONE IMPORTANT debate (*Ad Hoc* Political Committee set up by the General Assembly) in which I took part was on Korea. Having been India's representative in the Far Eastern Commission in Washington,* I was naturally interested in this debate. I asked Madame Pandit, the leader of the Indian Delegation to the General Assembly, what part we might play. She replied that there was no brief from our government on this question but, if I wished, I could take part in the debate — on my own responsibility. This was good enough for me. I prepared a brief and

*This Commission was established to keep a discreet watch on policy issues arising under US Supreme Commander General McArthur's administration of post-war Japan.

telegraphed it to Delhi for approval. But the debate started the very next day. (After the debate had ended and the Resolution approved, I returned to Washington to find a message from Delhi approving my brief!)

In September 1945, at the end of the war, Korea had been divided at the 38th parallel between US and Soviet Occupying Forces. In spite of the Cairo and Potsdam Declaration for a unified independent Korea, the two occupying powers could not agree on the procedure to form a government. Neither could they agree on a time-scale for withdrawing their respective forces. In May 1948, a People's Republic had been set up in North Korea, claiming jurisdiction over the whole country. In July of the same year, another government had been set up in South Korea, under President Syngham Rhee, also claiming jurisdiction of the country and supported by the US and China.

The matter came up before the UN General Assembly in 1949. This was against the wishes of the USSR. Having organized the nucleus of a communist party and a trained army, the USSR insisted that both the Occupying Forces should withdraw immediately and leave the people of North and South Korea to work out their own future. The US would not agree to this demand, since they had not been able to set up the same arrangements in their zone. Thus, the US called for the UN to debate the issue.

The purpose of the debate was to find a way to break this deadlock. I put forward some constructive suggestions of my own, several of which John Foster Dulles of the USA accepted and incorporated in his resolution. The Resolution provided that there should be a general election, not on a zonal basis, but for the whole country, on adult suffrage and by secret ballot; that immediately after the formation of a National Government, all military and semi-military bodies, not included in the national army, were to be disbanded; and finally, that the Occupying Forces should be withdrawn "as early as practicable" (this was a deviation from my suggestion of "within a definite time limit"). This Resolution was adopted after the East European delegations, led by Gromyko (USSR), had withdrawn from the meeting in protest.

The Korean debate is no longer of interest to the general public. But for me it was important because, for the first time, it brought me into direct contact with the turbulent politics of the Super-powers, giving me an insight into their thinking and an opportunity to observe them at play on the world chessboard in an atmosphere of Cold War. This experience was invaluable for my work in later years. If I may quote from the printed proceedings of the debate:

Mr Sen (India) proposed: "First a general election should be held, not on a zonal basis alone but on a national basis under the control of the UN temporary commission. That would be necessary to remove the political and moral barrier that had been created by the division of the country.

Secondly, the election should be held on the basis of adult suffrage without any political discrimination, and by secret ballot. That would facilitate a free election and would avoid any attempt to deny the vote to certain classes of people classified as undemocratic. Thirdly, the Assembly should meet, immediately after it had been elected, to form a national government. Fourthly, the national government, immediately after its formation, should constitute its own security forces and dissolve all military and semi-military formations not included in the national force. Lastly, a definite time-limit should be fixed for the withdrawal of the occupying troops, a matter of prime importance."

Dulles (USA): "the new text of paragraph 2 of the operative part (on the holding of general elections) ... including the suggestions ... of India ... that the elections should be held on the basis of adult suffrage and by secret ballot had been taken from the statements of the Indian representative. Paragraph 4(a) included a phrase "dissolve all military or semi-military formations not included therein ..." and was the precise phraseology suggested by the Indian representative. In paragraph 4(c), it was stated that the occupation forces should be withdrawn as early as practicable "because it was uncertain whether transportation would be available within that time."

The sequel to this whole matter is well known. The USSR withdrew its force late in 1949; the US departed in May 1950. The country remained divided at the 38th parallel — the North under General Kim II Sung, the South under President Syngham Rhee. No general election involving the whole of Korea ever materialized. The whole edifice was finally shattered when, in June 1950, the well-trained North Korean Army struck across the 38th parallel, starting the Korean War.

The 1949 Korean Debate attracted the star performers of the day to the United Nations — men like John Foster Dulles of the USA, André Gromyko of the USSR, Wellington Koo of China and Manuilsky of the Ukraine; at a lower level came Evatt of Australia, Couve de Murville of France and Sir Alexander Cadogan of the UK, ably assisted by Sir Hartley Shawcross, Britain's Attorney-General. I found it ironic that a newcomer and neophyte like myself was able to help break the Korean deadlock that existed. I shall always remember my encounter with Wellington Koo of China, a notable figure in the diplomatic world and Ambassador to the USA at the time. My proposals on the Korean question had raised quite a stir in the Chamber and both Hartley Shawcross (UK) and Dulles (USA) came to compliment me afterwards. At the end of the day, Wellington Koo asked me to travel with him to New York. At that time, the UN General Assembly was held at Lake Success, an hour's drive from the city. Throughout the journey, Wellington Koo pleaded with me not to press my proposals since the withdrawal of US forces from South Korea would seriously endanger the security of China. I said that I could not retract my proposals at that

My father

Dr. Radhakhrisnan, India's President and my life-long friend ever since our student days together at Calcutta University

President Tito held India's leaders in high esteem; there was a bond of mutual friendship between our two countries during my time as Ambassador to Yugoslavia

During my two terms as Ambassador in Italy (1950-51 and 1953-54), my wife Chiroprova and I immersed ourselves in the art, history and sights of the country

Prime Minister Nehru had an unrivalled knowledge of foreign affairs and he initiated many policies for India abroad, establishing diplomatic missions all over the world; Nehru is pictured here with his team of ambassadors for Europe and North America

With Mamoru Shigemitsu (right) Japan's Foreign Minister, and Dr. Radhakhrisnan

Madame Pandit (left) was the leader of the Indian Delegation to the United Nations General Assembly when the Korean Debate came up in 1949. My participation in this debate was my first exposure to the turbulent politics of the super powers

Swami Yoganada personified the true spirit of India. He was a great spiritual and wise leader, my acquaintance w him being all too brief – a few days at his Centre in Los Angeles

致 蔣總統之公開信

「皮之不存 毛將焉附」

一錄 蔣總統在發金圓券聲明中所引用之中國古諺一

詹士登國際公司伊德爾謹上

HARRY EDELL, JOHNSTON INTERNATIONAL
"A DIVISION OF THE JOHNSTON PUMP COMPANY, LOS ANGELES, CALIF."

(Above) The original letter in Chinese and (right) an English translation

An Open Letter to President Chiang Kai-shek:

"IF THE SKIN IS GONE THE HAIR WILL HAVE NOTHING TO GROW ON..."

Old Chinese Proverb quoted by President Chiang in his statement on issuance of new gold yuan

August 23, 1948

Your Excellency:

Permit me to extend my personal congratulations to you and your entire cabinet on the event of the birth of your new gold yuan currency policy.

The nations of the world whose interests lie in the uplifting of humanity will view this great forward action on the part of your government with genuine sympathy and understanding.

The soil of China upon which 360,000,000 loyal farmers barely eke out a living is also "as the skin is to the hair." Constant use of the same 210,000,000 acres which have been used for generations on extremely small plots of ground by these farmers will also leave no skin on the head of China and eventually no food.

Your currency reform is an excellent step in the right direction. May I respectfully suggest that your government add to this another reform, namely, the opportunity for these farmers to increase the food production--at present far from sufficient--by issuing land grants from the 100,000,000 acres still available for development.

China will then be on the threshold of tremendous prosperity due to the great potential buying power which will result therefrom.

American industry with its great resources and scientific methods stands ready to help China in this most worthy and great undertaking.

Democracy comes from the heart and Communism from the stomach.

With assurances of my highest esteem, I remain,

Respectfully,

Harry Edell
Johnston International
"A Division of the Johnston Pump Company, Los Angeles, Calif."

Whilst in China in 1948, investigating agriculture and food production, this open letter from an American businessman to Chiang Kai-shek appeared in all the local Shanghai newspapers, congratulating the President on certain actions and suggesting further reforms

Winter sports at Terminello, Italy

My diplomatic career had its lighter moments – such as being shown around the Hollywood Studios by Mr. and Mrs. Warner (left) and meeting some of the glamourous stars

stage of the debate and, in any case, they had found favour with at least one of the Occupying Powers. The fate of China was indeed in the balance. A year later, I happened to be in Peking at the head of a UN/FAO Mission. Mao's forces had already reached the Great Wall and I had missed the last bus! (I had to wait almost 30 years, until 1977, to visit Peking again: see *Postscript,* Chapter 27, page 288). I also remember seeing within the Forbidden City on that first visit a large number of students who had walked a thousand miles from Harbin University before Mao's advancing troops. Wellington Koo was a fine orator — there are stories of how he could take people by surprise with his superb command of the English language. I recall his suave personality and his style as a diplomat of the old school.

Apart from this important experience, I was also elected Chairman of a Committee set up by the Security Council to look into the rules for admission of new members, in order to facilitate the process. In dealing with this matter, I came to realize that nothing short of another world war, wiping out the institutions of the last, would allow the slightest change to be made in the Charter without opening a floodgate of controversy. It is to be remembered that the Charter was written when over half the world was under Colonial rule and UN membership only 65 nations (as against 151 at present); also, apart from the five Veto powers, there were others now who could legitimately claim the same privileged position and a country of 600 million people had the same one vote as a country with less than one hundred thousand.

My participation in ECOSOC debates as leader of the delegation of my country in 1949, and later in 1953 and 1954, also widened my experience in the problems of the United Nations. I still vividly recall some of the UN figures at that time: Vyshynsky (USSR), a powerful speaker whose torrential orations almost seemed to put his English interpreter into a trance, so that he would repeat even Vyshynsky's coughing as he went along; André Gromyko, one of the most brilliant men of the time, who could speak for hours without any notes, whose only physical movement during long speeches would be an occasional rolling of a pencil between his fingers; his fault was that he hardly raised his voice to stress a point — he spoke as if he was dictating to his secretary. Another person who left an impression was El Khoury (Syria). The Arab-Israeli war had then spilled over to the UN. El Khoury was the main spokesman for the Palestinians. I came to know him well and years later when on a visit to Damascus I visited him to pay my respects. He was in a darkened room — his eyesight had gone. He reminisced about those stirring days.

Another figure I recall was Bebler, a youthful representative of Yugoslavia, who had the nerve and audacity, born out of the new vitality experienced in his native country, to take on the whole world without a care. One lone figure who left a permanent impression on my mind was the Greek representative at the Political Committee. He was the Greek Foreign Minister, Politis, whose calm but impassioned and reasoned statements on the woes of his beleagured country (then under heavy Communist assault) made me think of his ancient orator-ancestor, Demosthenes.

At ECOSOC, a colourful figure was the Philippine representative. He seemed so fascinated by his own orations that, when he was finished speaking, he would gather up his papers and leave, giving the impression that whatever else was said thereafter would be an anti-climax. Charles Malik of Lebanon was a fine speaker and always made constructive suggestions. Among the Secretariat staff, apart from David Owen (Assistant Secretary for Economic Affairs) who was always worth listening to, the person I remember best is Mrs. Alva Myrdal, then head of the Social Affairs Division in the UN. I heard her speak only once but her presentation was clear and precise. Most impressive was her poise in that male-orientated world! Even today, we do not see such figures in the UN Secretariat, despite the advent of International Women's Year!

Visit of Prime Minister Nehru to the United States (1949)

I MENTION Nehru's visit to the USA under the theme of the Cold War because there were various aspects of the new India in which the United States was interested. Nehru was emerging as a world leader with a philosophy of his own which did not quite conform to expectations. The United States wanted to influence India and win her support in the Cold War. Direct contact between Nehru and US leaders in government, industry and education was regarded as essential.

Nehru came at the personal invitation of President Truman. The red carpet was laid out everywhere. The procession of outriders with screaming sirens, halted traffic and great fanfare became a feature wherever he went. No effort was spared to demonstrate the respect that the United States had for the India of Gandhi and Nehru.

I was entrusted by Madame Pandit to draw up his engagements and itinerary in consultation with all concerned. I accompanied him and Madame Pandit to many of the engagements — to the State Department for his call on the Secretary of State, Dean Acheson; to the Mayorial Reception in New York; to his speaking engagements at the Foreign Policy Association, the East and West Association and to the Institute of Pacific Relations in New York. I was one of the guests when Eisenhower, then President of Columbia University, held a reception at his residence in honour of Nehru. The display of gold-handled Japanese swords in exquisitely inlaid scabbards and other war trophies was most impressive. Wherever he went Nehru spoke without notes but his speeches were some of the most luminous that he had ever made. In all of these, his warmth, transparent sincerity and world outlook shone through. His visit was a great personal success.

But consistently he maintained that India must stand on her own feet in economic matters but was prepared to seek outside help in terms of mutual benefit. "We realize that self-help is the first condition of success for a nation, no less than for an individual. We are conscious that ours must be the primary effort and we shall seek succour from none to escape from any part of our own responsibility. But, though our economic potential is great, its conversion into finished wealth will need much technical and technological aid. We shall, therefore, gladly welcome such aid and cooperation on terms that are of mutual benefit."

On foreign policy he was less compromising. He clearly maintained India's policy of "non-alignment". His remarks on nuclear tests — this against the background of the news from the USSR — were not well taken. "Must this unhappy state (of a common fear for the future) persist and the power of science and wealth continue to be harnessed to the service of destruction? Every nation, great and small, has to answer this question and the greater the nation, the greater is its responsibility to find and to work for the right answer." These words, in fact, created some resentment and were regarded as "lecturing". Would he be neutral in all circumstances? No, he would not, but he would go no further than to say: "We are neither blind to reality nor do we propose to acquiesce in any challenge to man's freedom from whatever quarter it may come. Where freedom is menaced or justice threatened or where aggression takes place, we cannot be and shall not be neutral."

What Nehru said was clear and open. But, while he evoked respect as a person and as a leader of the Third World, he did not make the government, industry or the intellectuals in the United States either happy or comfortable. Nehru would remain neutral and non-aligned. India's economic philosophy would be socialistic, thereby restricting the scope of private industry from abroad. These were the impressions that Nehru left behind him in the United States.

It is my view that this visit, instead of bringing the two countries closer together as the United States had hoped, started a process of alienation which has not yet been wholly repaired. The immediate effects of this alienation could be seen in the debates on the Kashmir question which came up before the Security Council from time to time. The US Military Establishment also became more and more open in its preference for our neighbour in arms supply and build-up. I had to approach the State Department on more than one occasion over this matter. The "tilt" that President Nixon expressed years later was a culmination of this process.

About this time I was told that my posting as Ambassador to Italy and Yugoslavia, as a dual charge, was under contemplation. These countries

Treaty of London, signed in 1918, gave the city to Italy and the Italians occupied Trieste in November of that year. As part of Italy the city lost its old political importance but became a big industrial centre with shipbuilding, steel mills and oil refineries.

The Germans occupied Trieste in September 1943. In April 1945, Marshal Tito's troops after liberating Dalmatia and Istria entered Trieste and incorporated it into Yugoslavia. The Peace Treaty with Italy, signed in Paris in February 1947, decided that Trieste should be a small independent neutral state in view of its past history. It was to be called the Free Territory of Trieste, with its integrity and independence guaranteed by the UN Security Council. Temporarily it was to be divided into two zones, Zone A under British-American military administration and Zone B under the administration of Yugoslavia. It proved an unworkable arrangement from the very beginning and frictions grew.

On March 20 1948, the US, the UK and France proposed the inclusion of the whole territory in Italy. It was a decision without prior consultation with the parties concerned and naturally was wholly unacceptable to Yugoslavia. The matter came up before the Security Council and the debates were inconclusive. The administration of Zone A under US-UK military command continued as before. In 1952, I sent word to the British Commander that myself and my family would be passing through Trieste to go to Yugoslavia. He asked me to stay with him for a couple of days. He turned out an Honour Guard to welcome me. We spent two very happy days with him in his enchanting castle which rose sheer out of the sea. It was really a wonderful place, made for romance. During my stay I also visited the *Miramar* Castle built by Maximilian. Some of the rooms were as he had left them. There was an atmosphere of tragedy hanging over the whole castle. His downfall seemed to have happened only a short while before!

On October 8 1953, the US and the UK proposed the partition of the territory along the existing lines and handed over Zone A to Yugoslavia. This again was done without prior consultation with the parties concerned. It was at this point — October 1953 — that both Italy and Yugoslavia brought up their soldiers, tanks and artillery to the city border. Italy was going to occupy Zone A and Yugoslavia was going to keep the Italians out. The situation was tense. An imminent war was feared in Italy. I attended the Parliamentary debates to hear Pella explain the situation. I wanted to see the situation for myself. It was not easy to arrange to go first through the Italian position and having passed through the city under US and UK Command, to then go through the Yugoslav position. But it was arranged as I was Ambassador to both countries. I had the feeling that all this was really a melodramatic show with both sides eager to register their respective claims rather than really become embroiled in a full-scale war. I was wished *bon voyage* on both sides.

In February 1954, the US and the UK began negotiations in London with the two powers directly involved and hoped to find a final solution to the problem. Both Pella and Tito showed moderation and this open sore seemed at last to be healed for ever. Trieste is now a declining city having lost its strategic importance and some of its major industries are in recession.

As the time came for me to say goodbye to both Italy and Yugoslavia I felt sad. These two countries had enriched me enormously — Italy because of its past history as the centre of Western civilisation and its untold treasures in art and architecture, and Yugoslavia because of its recent struggle for identity and independence. I was specially sorry to take leave of Tito whom I held in such admiration. On my way back from Belgrade, my daughter, the only one of the family who was with me, wanted to have a last look at Dubrovnik. We started out in our car. On reaching Sarajevo, which was half-way, I learnt from the British Consul that the main road to Dubrovnik was closed. However, with my newly acquired Yugoslav spirit, I was not to be dissuaded and I did not want to disappoint my daughter. After local enquiries, I discovered that there was a track used by sheep and carts going through the mountains which I could try. We proceeded. It was a hair-raising experience. In some places the track was only the width of the car. One side was the mountain rock-face and the other, a sheer drop. Our Italian driver had worked as a driver-mechanic in the Abyssinian War. I have never known a finer driver. He struggled along without turning a hair. He said it reminded him of the Abyssinian mountains. We arrived in Dubrovnik all in one piece. After a day or two we set off again. Soon after we left Split, we discovered that a bridge spanning a wide river had collapsed and it was impossible to proceed. However, we could not turn back. On the map we saw a narrow road going up to Gospić and then running parallel to the coastal road. We took that road but before we got to Gospić all four wheels had punctures. Somehow we reached the place and got in touch with the local mayor. He was most obliging and said he would have the tyres repaired by next morning. We stayed the night in the waiting room of the local railway station. It was winter and freezing cold but we survived. Next morning we were on the road again and finally came down from the mountains at Senj to reach the coastal road. It was in its own way a rather exciting and colourful exit from a country that I could never forget.

Washington

I RETURNED to Washington after a lapse of two years and I did not find any striking change outwardly. I presented my credentials to President Truman. I renewed my contacts with the State Department and found several of my old friends still there. However, it did not take long to discover the change in the political climate that had since taken place. I have already referred to the reactions to Nehru's visit and his enunciation of India's non-alignment policy in the Cold War. The new Assistant Secretary of State for Asia and the Far East who had come from the Marines had not yet developed the *finesse* of a diplomat by Talleyrand's definition. He clearly showed his hand and he had no love for India. Incidentally, one of my early calls on the State Department was to see Admiral Nimitz who had just been appointed Mediator by the Security Council in the Kashmir dispute. He was wondering whether he could finish his work in three or four months. I was appalled at his optimism which he must have acquired from his spectacular successes in the Pacific War.

One of my main activities during this term of office was to go round the country to explain the thoughts and aspirations of the Indian people and Government. My address to the English Speaking Union in New York will give an idea of what I was trying to do. In spite of the disenchantment of the US Government and the political community generally about Gandhi, Nehru and the new India, the public were still interested and there was always a full house when I spoke.

Another matter to which I gave particular attention during my trips across the country was the question of the "brain drain" from India. I visited most of the premier universities and met the Indian students singly or collectively as well as the Presidents and Heads of the Faculties. At some places I was the house guest of the President. The Indian students and research scholars had established quite a high reputation for themselves over the years and they were very welcome. When we talk of "brain drain" we think specially of students and scholars who come for advanced studies and research in science and technology. The general complaint of the developing countries which now finds expression in international fora is that such students and scholars are often persuaded to stay in the West by material inducements such as higher salaries and a more lucrative career. Was this the whole truth? That was what I wanted to explore. The impression I gathered was that most of the students wanted to get back home and had a sense of service for their own country. In some cases they stayed abroad but not for the reasons generally ascribed. Several instances were cited by the Presidents and Professors where students went back home and then returned. They returned mostly because there were no openings for their special lines of research. They had specialised too much for the country to absorb them.

One President mentioned a case by name and said that this was the second time the student had gone home and, if he returned this time, he would personally go round for his naturalisation papers and give him a permanent appointment in his University. I wrote to my Government explaining the situation. I suggested that the matter be examined in depth and with objectivity and suitable steps taken to give the right incentives to the students to return. These incentives might include opening up new lines of research in the already established Research Centres or perhaps altogether new research stations depending on the importance of the research, widening the terms of Public Service appointments to encourage "over-qualified" candidates to apply and new criteria for university grants by the Central Government. I received no reply. The Government machinery apparently proved too slow or static to accommodate such new ideas.

I am tempted to make some general observations here. The approach of the developing countries appears to me to be a little too narrow in this matter. Instead of laying all the blame on the selfishness of the students or on the developed countries, developing countries have a part to play. There is the case of Dr. Khorana in India who had to go to the United States to carry on his basic research in human genetics and went on to win the Nobel Prize. He could not find the tools nor the other necessary minimum facilities in his own country. Need we now make a grievance in such a case? China provides another aspect of the same problem. During the last thirty years under Mao's policy of self-sufficiency and self-containment the practice developed of sending students to work in the fields irrespective of their merits or their intellectual promise. This had disastrous results for China's scientific development. (Those who developed China's atom bomb had had their education in the USA before Mao's regime.) China now is opening her doors. Thousands of promising students are being sent abroad to bring back to the country the latest knowledge of science and technology so that, by the year 2000, China will reach the level of the most advanced countries in the world. This should be an example to the rest of the developing countries. The question assumes particular significance in the context of the transfer of science and technology which is now high on the agenda of the East-West confrontation. Knowledge should not be limited by territorial boundaries. Advances in knowledge should be the legacy of all mankind. Today the developing countries want the transfer of science and technology from the West. By the same token they should not grudge the utilisation of talent which they themselves cannot utilise.

During the year McCarthyism was in full swing and Congress was busy arraigning people held in high esteem who were deemed to be tainted with communistic ideas and therefore were to be purged from society. It went to such lengths that even men like Marshall were not spared. The whole atmosphere was charged with fear. I was invited by the President of Colgate University to give an address. I had just come back from California where I had visited the President of Berkley University and learnt of the effects the persecution of intellectuals was having on his faculty members. In my formal address to Colgate University I gave expression to the dismay of the outside world at this sad spectacle in a country which had in the past been regarded as the home of liberty. I do not recollect but I must have used very harsh language for the next morning the address was carried by the press from coast to coast. On my return to Washington, I learned that my remarks had been objected to by a senior member of the Senate Foreign Relations Committee as interference in the internal affairs of the country deserving my withdrawal as *persona non grata*. However, the absurdity of this attitude must have been apparent to those responsible in Government for I heard no more about it.

This was the year of the Presidential election and the Primaries were on. I got caught up in it in a somewhat unconventional way. I happened to be at New York Airport to meet my teenage daughters arriving from Switzerland for their school vacation. A man came by to sell "I Like Ike" buttons. I picked up a few buttons from his plate and fixed them on the coat lapels of my daughters. It was a gesture of fun not of support for Eisenhower. A keen-eyed pressman noticed it and next morning it was in every paper — no doubt because Eisenhower was so popular. I got an angry cable from my Government a few days later wanting to know why I had indulged in such a gaffe. I thought silence was my best policy.

Then came the Republican Convention at Chicago. I was among the diplomats invited. to attend. I was particularly interested to see an American Presidential Convention. What I saw exceeded all my expectations. Though the delegates from the different States were seated in proper order, once the proceedings started it appeared more like a country fair with all kinds of side-shows to please the crowd. After each State cast its vote, there were parades with banners and supporters shouting and singing. Before the voting started General McArthur made a solemn entry in measured steps. The whole assembly stood up to give him an ovation. The ovation went on until I wondered if the proceedings were ever going to start. I expect the ovation was given as much to express their admiration for him personally as the great American warrior and Pro-Consul, as an act of disapproval of the Democrat Truman who had dismissed him from office. When I returned to Washington, a reporter came to interview me about my impressions of the Convention. His article next morning came out with the headline "Oh, Brother!" Those two words were indeed an apt comment.

I had an unforgettable experience in Los Angeles when I met Swami Yogananda at his Self-Realization Centre. He was a great spiritual leader whose teachings had attracted people from far and wide. I found in his Centre the peaceful atmosphere of an *ashram* such as we read of in our ancient books. The disciples, men and women dressed in *gerua*, chanting vedic hymns made me feel transported in spirit to the times of the great Guru Vashistha. Swami Yogananda personified the true spirit of the India we are so proud of but he was not altogether detached from the India of today. He welcomed me with great warmth as the first representative of independent India. His benediction enveloped me. I was with him for only a few days. He was not in good health but had me with him as often as possible. A large banquet was organized by the considerable Indian community of the West Coast to welcome me. Swamiji came all the way from his desert retreat. He told some of his close disciples that they must be prepared for his departure from this earth because he knew he would not be with them much longer. At the banquet he was seated on the podium between my wife and the presiding host. After I and some others had spoken he got up to speak. As he addressed us he did not show any signs of fatigue or distress but as he was saying his very last words he fell back from the podium onto the floor. In a moment his life was gone. I am not a deeply spiritual person. My life has been almost totally given to earthly pursuits. Here with Swami Yogananda, I had an experience which made my life infinitely richer and fuller. I had not realized then how widespread was his spiritual influence. Years later, I happened to be in Buenos Aires. A young man approached me with a book and a photograph in his hand. The book was a collection of Swamiji's writings and sermons and the photograph was of Swamiji sitting with me and talking. He wanted to know more of Swamiji's spiritual message. I had the same experience in a hotel in San Marino, Italy. A girl in her twenties raised similar questions about Swamiji holding the same book and the same photograph. It is a troubled world today. The younger generation is looking for some light to guide their way. This interest in Swamiji's teachings was an expression of that quest. The visit with Swamiji remains one of my most cherished memories.

Mexico

IN RETROSPECT, I do not think that even as a temporary arrangement it was a good idea for our Government to ask me to carry on as Ambassador to USA and at the same time accredit me to Mexico. This was done not for reasons of economy but no doubt for a lack of suitable personnel. It was an impossible arrangement. The United States as the centre of world power needed full-time attention, while Mexico as one of the most important countries in Latin America, in its own right, called for a similar measure of attention.

I presented my credentials to President Miguel Alemân in the Spring of 1952. This was the sixth and last year of his term. But what attracted my immediate attention was that the name of his successor in 1953 was already to be seen in neon signs all over the city. Ruiz Cortines was to be the next President. Coming from the United States and having had the experience of other democratic Constitutions, this naturally raised questions in my mind. How was it possible for a successor to be so firmly announced eight months before the term of the incumbent had ended?

I found the answer in the past history of the country. The Mexican Constitution of 1917 on which the present political structure was based reflected the end-result of ceaseless struggle between the Conservatives and the Liberals all through the past centuries — the one wanting to maintain the substance of the old order, specially with regard to land holdings and the rights of the church, and the other pressing for more liberal reforms with respect to both. The wars and revolutions which this struggle generated had been further intensified by outside powers — the British, the French and the North Americans — supporting the Conservatives or the Liberals to serve their own interests. The invitation to Maximilian by the French — Napoleon III wanted to protect his investments in Mexico and he went so far as to mount an invasion of Mexico by a French army — was but one aspect of the nature of the struggle that was going on. All this had finally ended in the revolution of 1910 which had brought out a strong political and economic nationalism which laid the foundation for the Constitution of 1917.

Though democratic in framework with a broad programme for reform to give a better deal to the agriculturists and labour, and with social welfare having precedence over individual rights to property, the Constitution did not prevent the political structure developing certain special characteristics. The most striking aspect of this was the emergence of a single monolithic party in 1936, a party composed of the three main sectors of the electorate — the agriculturists, labour and the general public. It was called the Party of the Mexican Revolution (PMR), renamed the Party of Revolutionary Institutions (PRI) in 1946. There were a few other smaller parties which remained outside but these wielded no influence. The Chief Executive of the Party was to be the President and as such enabled to wield powers which in a democratic system would not normally be vested in a President. For instance, the Mexican President was to have extensive control over Foreign Affairs and being the leader of the majority Party would have no conflict with the Legislature. This was a feature which specially distinguished the Mexican situation from its great northern neighbour. However, it was to be noted that, in spite of this privileged position, the Mexican President was expected to exercise his powers judiciously and with discretion so as not to appear to act as a Dictator. While the President was given the privilege of announcing his successor, it was the Party which made the choice. We were told that President Alemân, as also his successor Ruiz Cortines, were the choice of the popular sector of the Party.

Within the short time I was there I could get only a very superficial view of the country and the people. I came in contact more with the intellectuals than with the politicians and I found them very alert and articulate.

I was accompanied by my old teacher, Suniti Kumar Chatterjee. He had come to Washington just before I was to leave for Mexico. When I became aware of his interest in Mexican history and Mayan and Aztec culture, I had him inducted as my Honorary Cultural *Attaché* to Mexico. I valued his help in understanding what I saw. He spent busy days with me. He was one of India's finest linguistic scholars.

At the hotel where we stayed, a mural extending the length of the whole wall had been hidden from view by a canvas covering. I was told that the mural was by Diego Rivera and it had been screened from public view as it showed his too-open sympathy and admiration for the revolutionary leaders of the Soviet Union. The striking exuberance of the paintings of modern Mexicans, like Diego Rivera, José Clemente Orozco and others, were a reflection of the Mexican character which was bursting with life.

My wife and I did not have much time for travel but we did see a few places near Mexico City. We visited the Chapultepec Palace which Maximilian (June 1864) had remodelled. The whole tragic story came back to mind as I went from room to room in the palace which had been left undisturbed since his time. He left *Miramar* in Trieste with his young wife, Princess Carlotta, to seek his fortune in the New World. He plunged into the turgid and turbulent political currents of his adopted country. After only three years he had to face the firing squad on the Hill of Bells outside Querétaro. These facts are told in the history books. Though he had come at the request of the Conservatives his sympathies lay with the people and he insisted on carrying on the policies of the Liberal Juárez who had been exiled from the country. When in captivity, he was given a chance to escape but he refused. He could not "honourably desert his people" and he died bravely. It was the belief among some Indians that he was a re-incarnation of their fair-skinned bearded god Quetzalcoatl. While he was in prison Carlotta had gone to Europe to plead with Napoleon III at whose special bidding Maximilian had gone to Mexico. She also pleaded with the Pope, but her appeals failed. In the stress of the ordeal she lost her sanity and never recovered. After Maximilian's death she returned to her native country which was Belgium. She lived another 60 years and was mentally ill for all that time. Perhaps what sustained her through those long twilight years were the memories of her happy days at *Miramar*!

We had a boat ride in the canals of Xochimilco which reminded us of gondola rides in Venice. We visited the University with its new challenging architecture which expressed the character of the new Mexican civilisation. The students by their behaviour also seemed to show signs of the turmoil the country had gone through during the past centuries.

The Aztec Temples of Tenochtitlán stood as a solid and solemn testimony to the proud empire that fell to the rapacious advance of the Spanish *conquistadores* under Hernán Cortes at the beginning of the 15th century. (I visited Yucatan in the south to see the Mayan remains many years later; the Mayan tribes resisted the Spanish invaders for over a century and a half.) The Aztecs, however, did not die; they lived through the *Mestizos* by intermarriage with the Spaniards. The *Mestizos* are now regarded as the truest representatives of the Mexican nation, forming over 60 per cent of the total population. I learnt that no one could aspire to be President of Mexico unless he was a *Mestizo* thus representing in his person the fusion of the Spanish and the Indian heritage.

I came away with the feeling that here in this country a giant was waking after a long undisturbed sleep. The resources of the country, both human and material, were enormous. It needed a period of peace and tranquility to consolidate its strength. Its Constitution was quite radical with emphasis on reforms relating to land-holding, reforms with regard to the position of the church which had been a disturbing factor for centuries, a vindication of the rights of labour and the importance of social justice and welfare.

Today Mexico stands poised for a new future with the discovery of vast oil resources which properly husbanded and used could bring a new life to its people, not only the rich and well-off but also to the masses.

Japan

MY LAST DIPLOMATIC assignment was as Ambassador to Japan. It was a happy ending to this part of my career. The short visit I paid years before had left me with a desire to know Japan better. The Russo-Japanese war (1904-5) had provided occasions for our childhood games. The victory of Japan had been a source of pride in our part of the world. Families in Bengal gave the names of Japanese admirals and generals to their children or grandchildren. My own interest in the Japanese people had grown with the years. But during my previous visit I had not been able to penetrate the mask that the Japanese people usually wore in their contacts with foreigners — the mask which was probably the result of three centuries of isolation from the rest of the world. I was, therefore, happy that I was now to be given a chance to get to know them better.

The residence of the Indian Ambassador was once the property of one of the nobility — not a "Stately Home" but comfortable enough, with one part of the building in Japanese *tatami* style and the other part in Western style. It had a beautiful Japanese garden. It was situated near the Wasada University and every day one could see the students passing by. The Chancery was located on a floor of the NaiGai Building which overlooked the Imperial Palace with its imposing moat.

I was due to present my credentials to the Emperor and was driven to the Palace in an Imperial Coach with attendants from the Imperial Household. Passersby noting the approach of the Imperial Coach stopped and bowed as was the old custom. The ceremony at the presentation of credentials was simple. The Emperor was in a morning suit. There were no regal frills. The only decorations in the room were some Japanese paintings on the walls and some impressive bonzais in tubs. I later understood that this was not the original palace which had been burnt down during the war. The Emperor had insisted that he would not have it rebuilt until his people, who had suffered so much destruction in Tokyo, had been rehabilitated.

Nine years had passed since the end of the war. In the city, all the war debris had been cleared but large open spaces still showed where the fire bombs fell. However, there were already signs of revival of the economy. The communist take-over in China and the Korean war (1950) had been an incentive to this revival. These nine years had brought about fundamental changes. It was interesting to observe these changes taking place in a culture where the old habits and institutions continued to struggle below the surface.

The Supreme Commander laid down the lines for transforming Japan into a democratic nation with ideals and institutions like the West. This meant not only changing the course of their history but also transforming their character. The latter was more difficult.

The Japanese considered themselves a people descended from various *Kami* (gods) and prided themselves as having a superior culture. The Emperor was regarded as being in direct lineage from the highest Sun goddess and was to be treated as of divine origin. These ideas were previously inculcated and fostered as national policy. Shinto was adopted as the State religion in 1871 and was developed not so much as a religion but as a supra-religious national cult for mobilising the nation. The minds of the young were moulded through *Sushin* (moral teaching) in primary schools, comprising mainly the teaching of loyalty to the family, the State and the Emperor. The *Samurai* ethical code (*Bushida*) was adopted which required people to live a spartan life, accept pain and privation cheerfully and face death before dishonour. These were all integral parts of the national policy.

To make the Japanese a peace-loving democratic nation, all these basic elements of their life had to be changed by the Americans. That was how the Supreme Commander approached the matter. In his 1946 New Year Message to the nation, the Emperor was made to abjure his attributed divinity and the racial superiority claimed by the Japanese people. The new Constitution, which was drafted by the Supreme Commander and rushed through to become law in May 1947, formally abolished the absolute powers of the Emperor and left him as a symbol of national unity, with only ceremonial functions. That the Emperor did not suffer the fate of the German Emperor after the First World War, nor of those condemned after the Second, was we learned because of the personal insistence of the Supreme Commander. It was well-known that under the old Meiji Constitution which gave him absolute powers, the Emperor never exercised them himself but left them to the oligarchy around him. Therefore, it was only just that the Emperor should have been spared from the ignominy. However, if the idea was to make the Emperor, under the new dispensation, a constitutional monarch like in Britain, one wondered if, in actual practice, the position of the Japanese Emperor was not rather detached from the affairs of government. The British Queen, we are told, has Red Despatch Boxes daily from No. 10 Downing Street and periodic visits from the Prime Minister. Attempts made since by the Conservatives in Japan to strengthen the Emperor's position to some extent have foundered because of the Socialist opposition in the Diet.

Yet, it must take time to change the minds and hearts of a people imbued with the Emperor-cult through centuries. About the time I arrived, there was the Nijubashi (double-arched bridge) incident at the entrance to the Imperial Palace. Thousands had come from the countryside to pay homage to the Emperor on 3rd January and wish him a Happy New Year. There was a stampede, several people were crushed and many were injured. It was a daily spectacle to see groups of villagers — men and women from a whole village community — at the entrance to the Palace, going in or coming out in procession, with the leader holding up a banner to identify a group. The time-honoured practice of village communities from all over the country rendering voluntary service in the grounds of the Imperial Garden had thus continued. It had always been regarded as a great honour. The selection of the village groups was made by the Imperial Household in the past and now probably by the Prefecture. Respect for the Emperor was very strong among the peasants and it was the peasant families who had supplied the bulk of the formidable Japanese army.

By an order of the Supreme Commander, Shinto was disestablished as a State religion. All official affiliations were withdrawn and support from public funds discontinued. Yet Shinto still flourishes on private contributions. It is still the religion of the Imperial family and four shrines have been set aside for their household rites. Many shrines are dedicated to deities but some are also dedicated to historical figures such as Emperor Meiji and General Nogi who had rendered conspicuous service to the nation. I made it a point to spend a day at the Grand Shrine of Ise to witness some of the rites and realized that this supra-religious national cult was too deeply felt to be so easily obliterated. The recent controversy over the inclusion of some of the heroes of the war (condemned by the War Crimes Tribunal) for honour at the Shinto shrine of Ise, and Japan's Prime Minister's visit to the shrine on that occasion, demonstrates the feeling that still exists.

By another order, the Supreme Commander abolished the *Sushin* courses in primary schools. Teaching of Japanese history and geography was also temporarily suspended pending issue of new texts. Then the entire system of Japanese education came under review by a fully fledged US Education Commission. The Commission naturally recommended an entirely new structure based on US models to make the Japanese good Americans. The control of primary education was taken out of the hands of the Education Ministry and vested in Prefectural and Local Boards both elective and responsive to local public opinion. The Education Ministry was merely allowed to give advice when it was asked for by the Boards. In this way it was hoped that the young pupils would no longer be moulded according to one national pattern. It did not take long to discover that what was good for America was not necessarily good for Japan. The main complaint was that the new system undermined parental discipline, national loyalty and personal morals. I visited some schools of the new system and observed the new generation of boys and girls assuming more and more of the American image. I also made it a point to visit some small schools to see if the old atmosphere still prevailed. I remember stopping at a small roadside school on hearing the voices of children. The teacher who was a middle-aged lady asked the children to stand to show their respect for the guest. Then she picked up her baton to conduct a chorus which was sung with unusual gusto. What struck me most was the look in her eye — the dedicated look of one used to shaping the children according to the *Sushin* code.

In Japan I saw some aspects of a discipline of children which linger in my mind. Whenever boys and girls came out of a school, I used to notice that they went in a two-by-two formation. They would also board a bus or alight in good order — no scrambling, no shouting, no fighting. Once, on the roadside of a remote village, I saw a number of boys in school uniform crouching in a line, like sprinters waiting to spring forward at the pistol shot. They were waiting for their school bus! One extraordinary sight I will never forget was right in the centre of Tokyo traffic. I saw a procession of toddlers, barely able to walk, solemnly proceeding two by two. All the noisy traffic had come to a standstill. Only one or two teachers were guiding them. That was how the pre-war Japanese nation was made. Another noticeable feature of education of children was the practice of taking them out on bus trips all over the country to visit national shrines and monuments. Everywhere we went sightseeing, we saw schoolchildren in uniform, their faces alight with interest.

Apart from making the Japanese a democratic nation in the image of the Americans, the Japanese war machine had also to be levelled so as to make it impossible for Japan to rise as a military power again. Allied to this was the punishment of war criminals. The new Constitution dealt with the first and the War Crimes Tribunal with the second.

Article 9 of the Constitution stated "... the Japanese people forever renounce war as a sovereign right of the nation and the threat or use of force as a means of settling international disputes." Renunciation of war in these terms suited the mood of the nation. Never in all their history had the Japanese nation suffered such a defeat. Kublai Khan's invasions (1274 and 1281) had been halted with the help of *Kamikaze;* 'the divine wind' had scattered the invading ships. Since then none had dared to challenge Japan. The scale and nature of defeat this time had brought a genuine mood of soul-searching. Even as late as 1963, a Japanese representative at the World Food Congress in Washington prefaced his formal address by offering apologies for Japan's entry into war. But quite apart from a sense of guilt, there was another and more concrete reason which made the renunciation meaningful. It was the need to mobilise all her available resources to rebuild her shattered economy and enable her to take her place once again among the advanced industrial countries of the world. It was in this mood again that Japan agreed to the Treaty of Mutual Cooperation and Security of 1951, leaving the burden of Japan's security entirely to the United States. However, the uneasiness of some political sections in Japan with such total disarmament continued and during the years I was there I often heard this expressed. At the same time, developments in China, the Cold War and finally the Korean War (1950) made it necessary for the Allied Powers to revise their position. In 1950 a "police reserve", renamed in 1952 as a "National Safety Force" with naval units and finally called the "National Defence Force" with an air arm, was brought into existence.

The new political structure was to be reinforced by a new economic and social structure. Apart from the sweeping Land Reform which proved to be of enduring value in bringing about prosperity to the countryside, the other measures included dismantling of the *Zaibatsus,* which were regarded as potential instruments for the revival of the Japanese war machine, and far-reaching labour legislation, modelled on the New Deal ideas of the United States, to enable labour to contest the management for political and economic power. However, before the ink was dry, in view of the march of events in the world, there were second thoughts on some of these measures so hastily introduced. The *Zaibatsus,* with some minor modifications, were allowed to re-group themselves in order to ensure a more rapid revival of Japanese industry to withstand the tide of events on the Asian mainland. The application of the Labour laws which had acted as heady wine for industrial labour tasting freedom for the first time had to be restrained to be more in tune with the process of industrial recovery. The controversial aspects of the programme of reparations and the purging of war criminals, as well as certain measures of decentralisation, were quietly allowed to lapse or to be put aside.

On a different level, the occupation authorities did not lose sight of the position of the male member as the head of the family which had been such a potent force in inculcating the doctrine of Emperor-worship. The law of primogeniture was abolished to be replaced by equality of inheritance for all members of the family. In other countries, this kind of inheritance law has led to fragmentation of holdings and in a few generations has made the holdings uneconomic. How the problem has been tackled in Japan since the introduction of this law I do not know.

It was interesting for me to be in Japan when these changes were in a state of flux. How many of the old values which had given Japan her distinctive strength and character were being lost in this deluge and what of worth was being retained was a matter for the Japanese to judge. But, as I watched the scene unfolding, there were one or two aspects that I found worth noting.

One was the "family life" relationship between the workers and the management. I was visiting a big textile factory at Nagoya, an important industrial centre. A large proportion of the workers were young girls. I learnt that these girls were mostly recruited from the countryside at fifteen or sixteen years of age and were not only properly housed and looked after when they came to the factory but were paid enough wages to provide them with a dowry when they left to get married. Even their marriages were sometimes arranged by the management. This concept of "family life" was no doubt an extension of the same old value that had found expression in feudal times and in *Sushin* which was now rigorously excluded from the school curriculum. Later, I learnt that the practice of combining the technique of the modern assembly-line in industry with this kind of social relationship was still a feature of many large-scale industries in Japan.

The *Zaibatsus* have continued to project this concept of a united family for their workers on a larger canvas. I have seen a workers' song quoted which ran like this:

> "For the building of a new Japan
> Let's put our strength and mind together,
> Doing our best to promote production,
> Sending our goods to the people of the world,
> Endlessly and continuously,
> Like water gushing from a fountain,
> Grow, industry, grow, grow, grow!
> Harmony and sincerity! Matsushita Electricity!"

This would not be judged good poetry but certainly it was good business. Viewed against the background of conflict in the so-called free countries of the world between labour and management, the Japanese have a lesson to teach.

Another feature which particularly attracted my attention was the position cottage industries held in the total economy of the country. The cottage industries were involved not only in the traditional crafts where special skills were required but also in making parts for more recently developed and bigger industries, such as cameras, radios, television sets, etc. Small-scale industries in Japan have played a vital role in her industrial recovery. The system has the advantage of cheap rural labour some of which is on a part-time basis. Countries like India have something to learn here now that attention is being given to bringing the rural poor into the mainstream of development.

So much for the political post-war maelstrom in Japan! I must say I greatly enjoyed the colour and pageantry of the country. The beautiful handicrafts attracted my family's attention. My wife and daughter took regular lessons in Japanese flower arrangement and pottery and my daughter also studied woodcut painting. Some of their handicrafts are part of our household collection to remind us of those happy days. We were invited to innumerable parties of different kinds. We saw colourful ceremonies of communion with the spirits of the dead such as O-Bon (Festival of the Lanterns) and Higan (Paramita). We also saw folklore festivals such as the Tanabata. The floating of straw and bamboo boats with lights and flowers and offerings of rice reminded me of a similar ceremony I attended at the holy place of Hardwar in India many years before.

We explored the beauties of Japanese gardens in Tokyo and Kyoto. The symbolic composition of the hirs-niwa garden in the Ryoanji Temple, once seen, is carried through life. Learned Zen priests gave different Buddhist names to the fifteen rocks separated by raked white sand and linked religio-philosophic principles with them. The distinctive feature of Japanese architecture is that all the structures of the past were constructed of wood and the architecture was of columns and beams rather than of arches and vaults. The Buddhist temples at Nara and some of the Japanese castles, such as the Himaji Castle of Hideyoshi, were similarly constructed. All these were a feast to our eyes and we constantly visited them, often accompanied by guests from India.

Buddhism came to Japan through Korea in the sixth century. At first it was confined to the Court and intellectual circles but it gradually spread among the masses with the encouragement of the government of the Nara period. From time to time, depending on the balance of political forces within the country, it had to come to terms with Confusinism (concerned with public and private morality) and Shintoism (dealing with national and communal cults) and with the Bushido (Code of Conduct of the Samurai Warriors). This past history has given the Japanese a certain flexibility of approach towards their very distinctive religions. In some Japanese homes, I have seen a Buddhist family altar and a Shinto family shrine. The year 1955 saw the celebrations of the 2500th anniversary of Lord Buddha's birthday. I was invited to many functions and I addressed a gathering at Kyoto. The Association of Indian and Buddhist Studies also invited me to their functions. My wife, daughter and I joined in many festivities connected with Buddhist lore. Probably because of this liberal attitude of the Japanese people towards religion, the Burmese Prime Minister U Nu during a visit to Japan invited all the different schools of Buddhism to come together in ecumenical harmony.

I have to confess that in spite of the superb works of art and architecture, I did not find much of interest in Zen Buddhism which had been the source of inspiration for these arts. I was conscious that this was due to my own inadequacy and perhaps to my lack of sufficient intellectual curiosity for exploring the abstract paths of a religion which appeared so far removed from the simple original teachings of Lord Buddha.

The higher circles of Japanese society opened their doors to us with much generosity — though not literally, for entertainment seldom took place in their homes. The members of the Royal Family had accepted the change in their status and their fortunes with great dignity. The Royal Household had continued to entertain the Diplomatic Corps by their traditional calendar. The Emperor's Birthday was a special occasion for gathering. Among the sponsored sports, the duck-netting at Niihama (Chiba Prefecture), near Tokyo, and the cormorant fishing near Nagoya on the Nagara river were the most interesting. The latter was specially so. We stayed overnight at Nagoya and then the next evening we were taken out in small boats. A large flare was held aloft to attract the fish. Two or three trained cormorants were made ready to go hunting and their throats were tied by tape to prevent them swallowing the fish. As fish gathered under the flare the cormorants dived in and caught as many fish as their throats could hold. The cormorants were then pulled up by an attached string and relieved of their burden. The fish were served at our table the next day. I do not remember whether I enjoyed the fish but the thought of the ingenious cruelty of the sport could not be avoided.

Flower shows in Tokyo were important social events, comparable to Ascot in England. This was the occasion when ladies displayed the latest fashions. The flower shows were indeed a unique aspect of Japanese culture associated with Japanese gardens which had come down from the Nara, Heian and Kamakura periods and were designed to be enjoyed not only as views for contemplation but as microcosms to explore. The display of chrysanthemums during the autumn showed the exquisite artistry of the Japanese gardeners who were not satisfied with what nature alone would give; they trained the flowers to suit different moods.

Japanese musical tradition goes back to ancient times. Music from Korea and then from China during the T'ang period and dance pantomimes from India, Siam, Burma and Indo-China came to hold sway and continued till modern times. I saw a performance of such pantomimes at a school in Tokyo, including some of our own specialised Indian dances, which I found most impressive. Considering their past history I was surprised at the passion of modern Japanese youth for

Western classical music. We often had visits from Europe of well-known orchestras and conductors. My friend, Ichiiro Furokaki, head of Japanese Broadcasting (NHK) often arranged such performances in public halls in Tokyo. At these performances some of the finest instrumental music was played by Japanese girls. One of our diplomatic colleagues, Ambassador Leitner of Austria, was another enthusiast for music, a true ambassador of his country in that sense. I specially remember the visit of the famous conductor Karl Münschinger with his Vienna Philharmonic which the Ambassador arranged. The Italian Ambassador was also prominent in this field. He followed the great traditions of his country and brought in troupes to give performances of Italian operas, in particular of Puccini. This reminds me that when I was on my visit to Nagasaki, a city that has experienced such horror, I was shown the home of the original Cho-Chosan, Puccini's Madame Butterfly.

I travelled the length and breadth of the country. I visited Hiroshima and Nagasaki. The *Gaimusho* (Foreign Office) arranged my visit to these two cities. At Hiroshima I was met by a local official. I noticed one side of his face was completely burnt. I asked if he could show me some traces of what the atom bomb had left behind. He immediately changed the conversation and talked of the new industries that had since grown up and whether my country would be interested in some of the products. The memory was too painful for him. As I proceeded by train towards Nagasaki, at every stop I heard the beat of drums and chanting. It was Buddhist monks come to pay homage to the Ambassador of the country of Lord Buddha! I was rather flattered.

Nagasaki still shows signs of the devastation of the atom bomb. A museum has been set-up to house all the available material connected with the bombing. On the walls in one room hang the warning posters, signed by Truman and dropped from the American bombers, proclaiming doom for Japan if she did not surrender. The photographs of the victims of the bombing — some too ghastly to describe — remain a visual reminder of what the world could face in the future. I visited the hospital which received the first victims. By some miracle, it had escaped the radiation and damage from the blast. Two radiation cases were still there — a boy and a girl. They were schoolchildren in 1945. The girl had wounds on her thigh, knee and leg which, she said, flared up from time to time and compelled her to come to the hospital. As I entered, the boy sat up on his bed, knelt and bowed. He was suffering from acute leukemia. I left flowers for them. The boy died two days after my visit. The suffering was over at last for him — August 6, the day for all mankind to remember.

In 1956, Mamoru Shigemitsu, who had signed the surrender of Japan on board the battleship *Missouri,* was Japan's Foreign Minister. He was widely respected, a Stoic in the old Greek sense of the word, without a trace of bitterness in spite of all that he had gone through. He hobbled on crutches, often in pain, having lost a leg in a bomb blast. He was always very kind to me and whenever I went to see him, he prolonged the visit much to the embarrassment of the aides responsible for his engagements. I told him that I wanted to see some of old Japan and he immediately put the *Gaimusho* to work, arranging a trip to the island of Shikoku in the inland sea of Japan.

I had just returned from this trip when I got a message from New Delhi asking me to proceed to Rome to stand for election as Director-General of the Food and Agriculture Organization (FAO). It was a surprise because this was the very first intimation I had of such a possibility. I looked upon it as a fresh challenge with infinite possibilities and I was glad to respond. Two years was too short a time to get to know Japan and the Japanese people but I was grateful for this brief and happy interlude.

Part II

Freedom from Hunger

Dedicated to all those
men and women
who laboured with me
"to strive, to seek, to find
and not to yield"

Preface

MANY excellent books have been written on the early history of The Food and Agriculture Organization — the forces that brought FAO into being and the personalities that were at work. I do not wish, however, to attempt another history of FAO. What I wish to do is tell the story of FAO's Freedom from Hunger Campaign and describe my life with FAO around that campaign. I will recall some of my experiences during that period — the initiatives, the successes, the failures, the unaccomplished tasks, the seeds sown, which after nearly a decade appear to be surfacing. I will also examine, though rather superficially, some new ideas which are in circulation and which are relevant.

Unfortunately, during the course of my busy life — though that cannot really be an excuse — I have never kept any notes, nor did I put by documents for further reference. Clearly I am not of the race of diarists like Evelyn or Pepys. Much of what I write comes from memory. When my memory fails, as it frequently does with names, I am blessed with a walking memory by my side in the person of my wife! A four-volume diary of all my engagements at Headquarters and while visiting countries around the world was kept hour-by-hour by my Principal Secretary, Jessica Campbell, for all the eleven years she was with me. This has proved invaluable. And lately, I have been able to re-establish access to the FAO and UN Libraries for the reference documents I needed. My account or my views may sometimes be open to criticism for mistake of fact or judgement. As regards the latter, I make no apology nor ask for any indulgence. My main concern is to tell the story how, during my time, FAO transformed itself from essentially a technical UN Organization into a development agency. If this book raises lively discussion on some of the issues, that would be sufficient justification for this venture on my part.

It may be relevant to mention that these reminiscences, which cover many questions of topical interest, were written and mostly completed between January and April 1979. Developments since then will give a grandstand view to the readers who wish to judge the accuracy of the observations I have made on some of these questions. I have particularly in mind Food and the Olympic Games which I feel are used as pawns in international politics.

Chapter 10

The Food and Agriculture Organization (FAO)

My Election as Director-General (1956)

I SPENT a fortnight going round the island of Shikoku, living in picturesque Japanese Inns, being entertained to innumerable Geisha parties, visiting fish-breeding stations, joining in a traditional folk dance, watching pearl-divers and finally visiting Buddhist temples shown to me with much reverence because I came from the land of Lord Buddha.

I returned home pleasantly tired and was handed a telegram from my Government asking me to proceed to Rome to stand as a candidate for the post of Director-General of the FAO. This was indeed a surprise as I had been given no warning. On my way to Rome I stopped at New Delhi and was told that the UK had made an approach to Prime Minister Nehru to put up my name. After a succession of Americans in this FAO post, the UK thought that there should be a change and of all the possible candidates I had the best chance. I felt flattered, but knowing the formidable position the US held in FAO, I did not feel very hopeful. I arrived in Rome and was met by the Indian Ambassador and taken straight to the Terme di Caracalla to see an Opera. The gay open-air scene swept away a lot of the tension of impending battle.

As the item of election of Director-General came up, it became clear that the ultimate contest would be between me and the American candidate, Dr. Davies, a Professor from Harvard University. Elizalde (Philippines) and Cantos (Spain) had little support. Dr. Mansholt (Netherlands) was quite hopeful though the first two ballots did not support his position. After the first ballot in the morning, both Elizalde and Cantos were eliminated. The second ballot in the afternoon left Davies ahead but short of the majority required. Mansholt trailed behind me.

In the evening of the same day, Dr. Mansholt came to see me at my hotel to ask me to withdraw in his favour as he had been promised Indian support earlier that year when he had visited New Delhi. He mentioned Ajit Prasad Jain, India's Minister of Agriculture, as having given that promise. I said that clearly I had much better support as shown by the two ballots and, in any case, I was acting under the direct orders of Prime Minister Nehru.

The next day was a Sunday and there was no session of the Conference. At midnight when I had retired to bed, my telephone rang and I was asked if I would go up immediately to Villa Taverna, the residence of the US Ambassador, to see Mrs. Claire Booth-Luce, whom I had known as a colleague two years previously. When I arrived, I found Villa Taverna in darkness except for the hall-light and a lone doorman waiting to show me in. Mrs. Luce received me as an old friend and after an exchange of generalities, broached the subject of the election. She said that if I was prepared to acknowledge at the Conference that the withdrawal of the American candidate showed the desire of the United States to see a representative of a developing country for the first time assume this high international position, she would persuade Davies to withdraw. But first she would have to ask the permission of Dulles, Secretary of State, who was then on a visit to Paris. I agreed. She rang up Dulles. It should be remembered that the shadow of the Suez crisis was still hanging over the international scene.

In my Acceptance Speech I carried out my promise. The only regret I have is that I had to defeat Dr. Mansholt, a man of sterling character, straight like Nimrod. I kept in touch with him afterwards when he distinguished himself as a leader in the EEC.

Dr. Davies returned to Harvard. A few years later he was appointed Head of UNRWA (United Nations Relief and Works Agency for Palestine Refugees in the Near East). I had visited the Refugee Camps in Jordan in 1958 and remembered the look of suppressed anger and despair in the eyes of the growing generation of Palestinians. But it was left to Dr. Davies to bring to the notice of the authorities the flames which were building-up and would soon erupt into violent acts, including high-jackings and killings around the world.

This reminds me of my first visit to Jerusalem during that same year. The city had been divided between Jordan and Israel under UN auspices (for which Ralph Bunche got the Nobel Peace Prize). The line of demarcation had been drawn on the map but, when the ground surveyors went to work, they took the line literally, disregarding all other considerations. One result which I saw was a house divided between the two countries, one half in Jordan and the other in Israel, with the grandparents on one side and their son and his family on the other, probably able to communicate only under cover of darkness! Some years later, I visited another city, divided this time by a wall — Berlin.

Another item on the agenda of the Conference at which I was elected was the problem of staff morale over which Dr. Cardon, Director-General, had resigned, making the election necessary. Dr. Cardon, whose life's training had been more as a scientist than an administrator, had apparently found the stresses and strains of living in an international community compressed within the walls of one House both unfamiliar and unmanageable. He was a sensitive person and when some extreme cases of breakdown of staff morale occurred — an interpreter was found dead in his booth suspected of having committed suicide — he had a nervous prostration himself leading to his resignation. I used to visit him at his home in Washington whenever I had occasion to be there. He lived quietly. Painting was his hobby, a talent he had discovered only after his retirement. Staff morale was not, however, a problem which daunted me. In my Acceptance Speech, I referred to this problem and I said I did not think that anything was really wrong in this area which, with the cooperation of Governments and staff, could not easily be brought under control.

Transfer of FAO Headquarters from Washington to Rome

THE TRANSFER of FAO to Rome in 1951 was a dramatic affair. Since I took an important part in its transfer I shall try and recapture some of that drama. Since its creation, FAO had been located in rented buildings in Washington. A statutory Rule had been adopted to say the seat "shall be" at the same place as the headquarters of the UN. The FAO Conference, however, had had offers from Denmark, Switzerland, USA (four alternative sites including the University of Maryland) and the UN (for part of the permanent headquarters in New York City).

Those who represented the USA were naturally anxious to keep FAO within the country. They felt that as they had played a large part in creating and sustaining FAO, both technically and financially, and being the most advanced agricultural country in the world with all the necessary expertise they had special claims. The US Delegation went to the unusual length of inviting President Truman to open the Conference just to plead their cause. In the Secretariat, the Director-General, Norris Dodd, and almost all his staff who did not want to be uprooted were against any move.

As Chairman of Commission III, I had to deal with the matter in the first instance. My own view was that Rome would be the best choice being midway for FAO countries both geographically and from the point of view of air communication. Denmark had offered the Elsinore Castle immortalised by Shakespeare's *Hamlet.* As a lover of Shakespeare, the lines "To be, or not to be . . . to die: to sleep, perchance to dream . . ." came to my mind to dissuade me. I wanted FAO to be more decisive in the pursuit of its objectives! I resisted pressures from all sides. Charles Brannan, US Secretary of Agriculture, who presided over the Conference, came up one day to tell me that Sir Feroze Khan Noon, Delegate of Pakistan, had agreed with the US and why would India not agree? I permitted a delegation of the US Association For the Advancement of Coloured People to appear before the Commission to plead against the University of Maryland on grounds of social discrimination. At the final Plenary Session which considered the recommendation of Commission III to transfer the seat to Rome, it was a prolonged exchange between me and the Legal Officer of FAO who at Dodd's behest was trying to prove the illegality of such a transfer.

The proceedings lasted the whole day and into the night. The resolution came to a vote around midnight and we won by a single vote. The printed FAO Report mentions a majority of two votes, probably because the vote of the unwilling Chairman, Charles Brannan, had to be counted on the side of the majority. All through the prolonged debate the Italian Delegation who happened to be sitting next to me as the seating was arranged in alphabetic order observed a discreet silence. The Italian Delegation was composed of Segni, who later became President of Italy, Emilio Colombo, now Italy's Foreign Minister and Professor Papi, who represented Italy in FAO for over 30 years. The main delegations on our side were the UK and France. Though young Harold Wilson, later to be Prime Minister, was the leader of the UK Delegation, I do not recollect his presence at any time during the Conference. Sir Donald Vanderpeer was an able substitute. Michel Cepède, then an exuberant young man, later to become Chairman of the Council, spoke for France. When I returned home long past midnight, I started to offer apologies to my wife, but I discovered that she had been there all through the evening session to watch what we agreed was quite a fight! After the transfer decision, the question arose of how to meet the extra expenses for the move. I suggested that the USA in their generosity might help us out! In making this suggestion, I had of course realized the irony. Two years later when I returned to Washington in my capacity as India's Ambassador, I had occasion to visit the office of the *Washington Post* and, while going round, the Editor pulled out from a drawer marked "S", an extract of the paper's reaction to my impudent suggestion!

The move did not please our unwilling staff. I was told that the ship carrying many of them ran into a heavy storm in mid-Atlantic and they considered it a warning omen to turn back! But they survived the move.

The First FAO Conference in Rome (1951) after the transfer was held in the picturesque Villa Borghese. There was, as yet, no Conference Hall to the FAO building. We insisted that the Hall must be completed within a year — an almost impossible task by normal building standards. But the Hall was completed within that time. The Italians are expert builders, whether it be roads or monuments, a talent they inherited from their Roman forebears.

Chapter 11

Internal Workings

Administration and Staff Management

ON ASSUMING CHARGE in November 1956, my first priority was the staff problem. Staff morale was vitally important. Even in those days, the staff at headquarters alone numbered more than 3,000, recruited from many countries, with different cultural backgrounds, speaking different languages, all under one roof. How to establish proper communication with them, giving them a feeling of unity was the problem. I tried to keep my doors open to anyone with any grievance or any representation to make. I did not allow my other preoccupations to stand in the way of easy access. I received the members of the Staff Council whenever they had any matter to bring up, encouraging a free exchange between us.

But apart from these well-established principles for a healthy administration, I thought it necessary and desirable that all members of the staff should be given a feeling of participation in the great mission that FAO was entrusted with. I made it a point to call them together in a general assembly after every FAO Conference to explain how the Conference had gone, what important issues had come up and what important decisions had been taken which could affect the Organization's future. I also kept them informed of any new initiatives of importance such as the World Food Congress to be held in Washington.

My wife and I took a genuine interest in personal problems of staff members. In fact, for the first time, I appointed an officer — a lady, a qualified psychiatrist, trained in dealing with problems usually encountered in an international staff far away from home. My wife and I also made it a point to attend and take part in social and athletic gatherings of staff members whenever possible.

My relations with the staff were not of the employer-employee kind, but of genuine partnership. I tried to cultivate the feeling that we all belonged to the same family and I wanted to show that I was interested in the welfare of every member irrespective of status. Some of my best friends (though some of them have died) were members of the General Staff. One I specially remember was Signor Ricciardi who had charge of the Communications system and he was an expert in electronic equipment. A finer, sweeter man I have not met.

Here I may recall one of the best compliments I received from the General Staff. Several years after I had left, a strike was called by the General Staff. When one of the militant staff representatives was confronted with the question why the staff had not gone on strike in my time when the issues raised were the same, the reply, as reported to me by a senior officer whom I met in the UN in New York, was "Oh, no! we would not go on strike in Dr. Sen's time!" This does not mean that I was soft when real breaches of discipline occurred. On one occasion of unrest fomented by a professional agitator on the staff, I heard that he had used violent language against the administration while addressing a staff gathering. I issued orders dismissing him on the spot, even though I knew that such action would find a sympathetic hearing in the Geneva Tribunal. Next morning I called the staff to the Conference Hall and told them, addressing myself especially to the difficult members of the Staff Council who were sitting on the front bench, that anyone following the example of the man I had dismissed would suffer the same fate. There was no further agitation of this kind until after my retirement in 1967.

I would like to record my debt to Frank Weisl, Assistant Director-General in Charge of Administration, on whose honesty, integrity and patient understanding I greatly relied. In that difficult position he was not always popular but he was universally respected.

My Senior Staff

ONE LESSON I had learnt from my career in the Indian Civil Service (ICS) which I carried with me to FAO was that the key to good administration lay in delegation of responsibility — to each according to his status and capacity. Among the senior officers, I had to do some initial sorting out to find the good thinkers, the good analysts, the experts in personnel matters and so on. But first, I had to look for a personal secretary who would have to carry a lot of my workload. I sought out Miss Jessica Campbell who had worked as my Secretary when I was in Washington in a different capacity. She remained with me all of the eleven years I was in FAO. I can never repay the debt I owe her for such devoted service. I also had a highly qualified Cabinet of ADGs with various talents — a rare draughtsman, a super-bureaucrat, an outstanding economist, a man of shining integrity. I would like to mention one of my colleagues by name — Dr. Mordecai Ezekiel. He was one of the finest economic brains I had the good fortune to come by. As a young man, he had distinguished himself and had become associated with Roosevelt's New Deal. A very humble man, with a charming wife of high culture whose friendship we still enjoy, he was indeed one of my main supports in the new initiatives I took in FAO — the Freedom from Hunger Campaign (FFHC) and the World Food Programme WFP, to name only two. His death was a deep personal grief for me and my wife. To get the best out of this talented team, I gave them the fullest freedom to think and to act but at the same time holding the reins in my own hands.

One way I did this was through a "Morning Meeting" every day with my ADGs. The ADGs would tell me their problems so I was kept informed and also was able to give advice if necessary. This helped to create a team spirit which proved invaluable. I had also inherited the PPB (Programme and Policy Board), a high-sounding name for meeting with the Directors and other senior officers. I used to have that meeting regularly, once every fortnight.

As I look back, I feel that I was fortunate in having such a group of talented and devoted colleagues. Any Director-General will have both successes and failures. For the successes, these colleagues of mine must be given, their due share. As I have said, a UN Organization provides a meeting ground for people of many lands and many cultures. This was an experience I specially valued. I came across men and women whom I came to respect and admire, whose friendship both my wife and I still enjoy though we seldom meet. One or two persons are always in my mind.

The first and foremost is Frank McDougall. It was in 1942 that he met President Roosevelt and outlined to him the idea of an international agency to deal with world problems of food production and distribution; when Roosevelt accepted the idea, McDougall sent that famous telegram to his friends in England, quoting Bishop Latimer who, about to be burnt at the stake, said "We have this day lit a candle . . ." At the Hot Springs Conference in 1943, he was a member of the Australian Delegation. He joined FAO as Counsellor and then as Special Assistant to the Director-General, in which capacity he represented FAO at the United Nations and ECOSOC. I first met him in Washington in 1947. I remember he took me to lunch to ask me what ideas I had about the future work of FAO. I doubt if I was able to throw much light on the subject. When I came to FAO, McDougall was still there. He died soon after. When it fell to me to propose a suitable memorial to this great friend of humanity, my suggestion, which was accepted, was for the establishment of a *McDougall Memorial Lecture* to be delivered by a person of world renown on problems of food and agriculture at the opening of every biennial conference. This Lecture has proved to be one of FAO's most valuable institutions. Certainly the first three were outstanding. Professor Arnold Toynbee was the first. I specially requested him to give his views on the population problem. When he came to this point in his address, the Vatican representatives sitting in the front walked out in protest. For the second I invited Mr. John D. Rockefeller. Again, I asked for his views on the same problem. He was even less diplomatic than Toynbee, but this time the Vatican representatives listened quietly and there were no incidents. The third Lecture was given by President Nyerere who spoke as a representative of the Third World. His address remains a high-water mark in FAO's history.

I also wish to mention Kay Kesteven (Australia). He was a well-known authority in his own field which was Animal Health and Diseases. Travelling in Australia, I made it a point to visit his college — where his name is inscribed on the top of the Honours Board. Apart from his professional distinction, what I admired in him most was the strength of his character. Towards the end of his career in FAO he suffered from an infection and one of his legs had to be amputated. When he came back from the hospital to resume his duties he appeared as cheerful as ever even though he had been told that he might lost the other leg.

I would like to add here that once I had their loyalty, I tried to protect my officers as far as I could. In one case soon after I had appointed an officer to a high post, I started getting telegrams from different countries in Europe accusing him of having been an associate of someone who had been found guilty of selling military secrets to an enemy country. Day after day, the telegrams kept piling up on my table until I called in Interpol. I heard no more about this and there were no more telegrams.

In another case an officer who had been appointed on merit but more so at the special request of a Member Government was later accused by the same Government of having been a member of an undesirable party. I said that I had no concern with the politics of any country. In neither case did I inform the officers of what was happening.

Politics and FAO

THE FACT that in 1945 international discussions at San Francisco for the creation of the United Nations and at Quebec for the creation of FAO were going on simultaneously showed that the founding fathers considered it was important for food to be kept out of politics as far as possible. The UN Charter provided the kind of relationship they envisaged with FAO and other Specialised Agencies. Under Articles 57 and 63 of the Charter, "coordination" was to be through "consultation with and recommendations to" such Agencies.

Having gone through the various manifestations of the problem of food in India, I came to FAO convinced that it should be protected at all costs, against incursion of politics not consistent with its basic role. I would like to think that on the whole I succeeded in doing so. But first of all,I had to condition myself to be a true international Civil Servant. This was specially important because for the previous ten years I had represented my country's interests as Ambassador to some of the countries whose goodwill India wished to develop, such as the USA, Japan and Yugoslavia. I also had to exorcise my negative feelings about those countries with whom India had had conflict in recent years.

I had known most of the political leaders of Pakistan quite intimately. I remember my visit with President Iskandor Meerza who was a colleague when India and Pakistan were one country. He asked me to come back and go hunting with him. Within three months he was overthrown in a military coup. I had not known General Ayub Khan before he became President but later he was one of my closest supporters. During the murky days of the 1967 election of the Director-General — the only period of my life with FAO that I would like to forget — when my nomination was voted down, it was the Delegate from Pakistan who was the main spokesman for those who wanted me to stay on.

My first contacts with Portugal were through the delegates who came to the FAO Conference. They brought me an invitation from their Government to visit their country. I particularly remember that visit. I paid a courtesy call to Prime Minister Salazar at his villa just outside Lisbon. It was a modest villa with no display of guards outside and a lone doorman ushered me in. Salazar was originally a University professor. He appeared just like what a learned Professor would be: quiet, slow-speaking and a low profile. We discussed many things about what FAO was doing. He was particularly interested in India's experiences in the economics of multi-purpose River Valley projects because Portugal was then in the process of building one in Africa. When we finished he came down with me and opened the door of my car, waiting until I departed. Was this the Great Dictator? Of course he was, but not in appearance not in his manner. While going through the Foreign Office and the Ministry of Agriculture, I had not noticed a single photograph of

Salazar anywhere such as would be ubiquitous in all "democratic" countries.

Immediately after assuming office I instituted a Monthly Letter to the Ministers of Agriculture of our Member Countries. I will give the main reason which lay behind these Monthly Letters.

"As Director-General", I said in my opening letter, "I am responsible for carrying out the policies laid down by the member governments, and at the same time for keeping them informed of the trends and developments. If the working of the Organization is to reflect from day to day the thinking and needs of the member countries, it is necessary for the Director-General to have as close a touch with them as possible. The purpose of this letter is to establish this sort of contact — to tell, every month, by direct approach, those who are at the helm of the administration of food and agriculture, what problems are arising and how they are proposed to be dealt with, what subjects are proposed for discussion by the governing bodies and in committees, and generally to try and keep them informed of important happenings." The reason why I decided to make the Ministers of Agriculture and not the Foreign Ministers my normal contacts was to keep the Organization out of politics. At the same time, sufficient extra copies were sent to the Agriculture Ministers to distribute to their colleagues as they thought fit. I recall that India's Home Minister, Govinda Vallabh Panth, one of the towering figures in Indian politics besides Gandhi and Nehru, was a regular reader of my letters and used to discuss them with me whenever I happened to be in New Delhi.

At the African Regional Conference in 1962, the Minister of Agriculture of the host country, Tunisia, brought up the question of eligibility of Portugal as a full member of the Regional Conference in view of Portugal's colonial policy. For two days the debate raged inconclusively. I took the view that the eligibility of a country for membership was not a matter for a Regional Conference but for the Plenipotentiary FAO Biennial Conference itself, and I warned that if the Tunisian delegate persisted in his argument, I would have no other alternative than to call off the Regional Conference altogether. I then sought an interview with President Bourghiba to explain my point of view. He readily agreed with me and immediately asked his Minister to give way. I took that position on entirely constitutional grounds in order to protect FAO from politics.

While on politics, I may observe that the decision to hold the World Food Conference in 1974 under UN auspices was certainly a departure from the idea of the founding fathers to keep FAO above politics. Perhaps it would be correct to say that more than any other Specialised Agency, FAO has been able to keep out of political controversies in its Conference and Council sessions. In recent years much credit is due to the present Director-General, Dr. Edouard Saouma.

When I think of politics and politicians of the present day, I cannot help recalling what I read about "rulers" in Plato's *Republic* in my student days. "To ensure honesty of those public servants, there must be no private property among them. They are to own everything in common, take their meals in public dining rooms, and sleep together in barracks. Having no personal interests, these rulers will be above bribery and have but a single ambition — to establish and to perpetuate justice among men." But that most comprehensive of the world's thinkers, as Emerson called him, knew that this could only happen in Utopia!

Chapter 12

Shaping of an Idea

The Freedom from Hunger Campaign (FFHC)

WILL THERE EVER emerge a world where there will be no degrading poverty, where all will have an equal opportunity for a better life? This was always my lifelong dream. In my different capacities as a public servant, I had seen and had to cope with famines, floods and other natural calamities affecting millions of people. As India's Director-General of Food during war-time (1943-46) my primary task was to ensure equitable distribution of scarce food supplies for one-sixth of the world's population. All my life I had been in the midst of hunger and poverty in all its stark reality.

The discussions at the Hot Springs, Quebec and Copenhagen Conferences had opened a new chapter in international solidarity and I had followed these discussions closely. But what gave me most encouragement was the visit of ex-President Herbert Hoover in 1946 as President Roosevelt's Personal Representative who was travelling round the world to assess the food scarcity situation. I had the responsibility to brief him about conditions in India and I took him round the country for twelve days. He talked freely to me and I was an eager listener. His great humanism and his unparalleled experience ever since World War I had left a deep impression on my mind. I recall some of the distinguished men and women he brought with him in his party. Maurice Pate, who became my friend and, as head of UNICEF, maintained the closest working relations with me in FAO; Margaret Mitchell, author of the famous book *Gone With the Wind* who died soon after her return in a street accident; and Margaret Bourke-White, the *Life* magazine photographer with her glistening, platinum-blonde hair.

The idea of mounting a World Campaign against Hunger was on my mind when I took over as DG at FAO. At the Summer Session of ECOSOC in 1958 I had an opportunity for the first time to raise the problem. In my address, I set out the main objectives of the campaign I had in mind: to attract worldwide attention to the problem; to secure the participation and cooperation of all concerned; to achieve a degree of enthusiasm and anticipation which would result in more effective national and international action; and in the process, establish a higher level of mutually profitable world trade to help raise the prosperity of both developed and developing countries. But how was such a campaign to be mounted? The only model of a world campaign I had before me was the 1957/58 International Geophysical Year which ran for eighteen months. Discussion of the IGY started in 1950. The Special Committee set up by the International Council for Scientific Unions started work in 1953. There was a full four-year period of preparation before the "Year" started on July 1, 1957. Following this precedent, I knew that a build-up period would be necessary before the programme was launched and eventually hold a campaign "Year" which would summarise the results, conclusions and agreements on an integrated programme of action at provincial, national and international levels.

I felt, somehow, that I had not quite got the answer in the proposals I had developed. The problems of poverty, hunger and malnutrition were so vast in scale and so serious in character that no one year's endeavour, however intense, could in itself bring solutions any nearer. I therefore told ECOSOC that our main aim should be to heighten awareness in the world and thus improve the foundations for effective and accelerated action which would hopefully continue for the future.

My proposal before ECOSOC evoked response from various parts of the world. Three months later I spoke to the FAO Council and I found my ideas had clarified. The campaign would be of an informational and promotional character with the expectation that a climate of public opinion would be created leading to intensification of action programmes both at the national and international levels. The Campaign — the idea of a sustained campaign had begun to take shape — should culminate in a World Food Congress that would sum up the main lessons and conclusions. I suggested 1963 as the date for the Congress because it coincided with the twentieth anniversary of the Hot Springs Conference which gave birth to the FAO.

FAO Conference (1959): Resolution on the Campaign

WHEN THE CONFERENCE came at the end of 1959 the character and scope of the Freedom from Hunger Campaign had taken full shape. The Conference Report and the Resolution adopted faithfully reflect all that I had proposed.*

When I opened the Campaign, I felt I could best give expression to mounting international concern by quoting one of my favourite poets, John Donne. I used the words "One man's hunger is every man's hunger — one man's freedom from hunger is neither a free nor a secure freedom until all men are free from hunger." These words were taken up as a key to the entire campaign.

In the Resolution, the Conference emphasised that the Campaign should be "under the leadership and general coordination of FAO" with invitations to participate to Member Countries of FAO, UN and UN Specialised Agencies and these Agencies themselves, international non-governmental organizations, religious groups, individuals and private organizations within Member Countries. The leadership of FAO in the Campaign was emphasised whenever the matter came up before the ensuing Conferences until I left at the end of 1967.

I interpreted this Resolution to mean that the leadership would have to be assumed primarily by the DG and that is what I did from the very beginning. Since this was intended "to promote a spirit of re-dedication of the Organization and its members to the achievement of the objectives of the Organization", I felt that it needed the involvement of the whole Organization with the DG at the centre personally controlling and directing all the activities bearing on the Campaign. In expressing the idea of a continuing Campaign, I had explained my understanding in these terms: "one, of course, realizes that a radical transformation of economic conditions cannot be achieved in a few years. What we can hope for is the generation of a tempo of development which may break the cycle of stagnation and lead to self-sustaining growth. This campaign is not intended to be a special new programme in itself, but only to aid and intensify our work in FAO and at the same time the work of member governments as well as of the vast masses of the people who depend on agriculture for their livelihood."

I found I could not carry the entire burden myself. I therefore created the office of a "Co-ordinator" within my personal Secretariat under my direct orders. The person I selected was Charles Weitz whom I had met in Ankara as UN Representative. His responsibility was mainly to keep in touch with the non-governmental organizations (NGOs) and groups, especially Church groups, undertaking direct responsibility for their

*See Appendix I on Page 305.

programmes. I attended and addressed most of the conferences of European FFHC Committees and representatives of NGOs, Voluntary Agencies and International Catholic Congresses. This I did as much to learn from and keep in personal touch with the leaders as to guide them in forming an effective coherent part of the total effort.

The World Food Survey (1962)

I REALIZED that one of the first questions that would be asked in relation to a campaign of this nature would be the extent of hunger and malnutrition in the world at the present time.

I therefore set in motion a World Food Survey as one of my first campaign initiatives. The survey took nearly two years to complete. Although it had been generally recognised for a long time that a considerable section of the world population was inadequately nourished, estimates of the extent of malnourishment had varied considerably. There was no agreement on the statistical yardsticks and this was further complicated by a lack of reliable data on food consumption for the entire world. The first such survey (1946) was more or less an extension of the work of John Boyd-Orr, which had been done for the British people (*Food, Health and Income*). Because there were serious gaps in statistical information, much of the material used was in the nature of intelligent guesswork. The second survey (1952) was somewhat more reliable as more realistic standards for calorie requirements could be established which took into account such factors as age, sex, body weight and average physical activity in differing environmental temperatures.

Ten years had passed since that survey had been done and during this time, not only had the impact of population growth revived Malthusian fears and anxieties about the adequacy of food supplies, but also new and significant data on food consumption patterns had become available. The Freedom from Hunger Campaign made it imperative to reappraise the situation. Such was the background of this important survey on hunger and future needs. Applying the FAO calorie standards, the survey showed that the calorie supplies in Europe, North America and Oceania exceeded the requirements by 20 per cent. In the Near East, Africa and Latin America, they were about equal to requirements. In the Far East the supply fell short by 11 per cent. (2,030 against 2,300 minimum required). The study showed that this gap was shared by one-fifth to one-fourth of the population of the region. Since this region comprised one-half of the total world population, it was clear that a conservative estimate of the undernourished (hungry people in the world) would be between 10 and 15 per cent and in actual numbers between 200 and 300 million people. As malnutrition refers to the qualitative aspects of the diet, the broad conclusion was that between one third and one half of the current world population suffered from malnutrition.

The survey was based on food-balance-sheets data for over 80 countries covering some 95 per cent of the world population. It drew upon food consumption and dietary surveys conducted in various parts of the world and introduced new statistical techniques in the study of food supplies and needs. It provided a concrete basis for our information and was quoted by world leaders for their appeals in support of the Freeedom from Hunger Campaign.

In order to close the hunger gap and substantially reduce the number of malnourished, the study also provided an estimate of the amount of food needed for the world's growing population. On the UN medium-term forecast, the food supplies would have to be doubled by 1980 and tripled by the end of the century. Thus the survey provided a basis for the formulation of future policies at national and international levels. Since then, the fourth World Food Survey (conducted in 1978, with more sophisticated statistical tools) calculated that the number of undernourished people amounts to 496 million. It will be seen that this is very close to the estimates worked out in my time.

A constant critic of our work was Dr. Colin Clark, a Professor at Oxford University. He was reported as saying that Dr. Sukhatme, who was the main architect of the survey, had merely acted as His Master's Voice, tailoring his figures to suit Sen's Campaign. His theory was that the food resources available were sufficient not only to feed the present world population but double that figure; that there was no need of birth control, that the Campaign was, in essence, nonsense. A few months after the publication of the survey, I was invited to address a gathering at London University. I was warned that among a distinguished group of intellectuals from both Oxford and Cambridge Dr. Colin Clark would be present. I spoke at length on our approach to the whole problem of world hunger and gave the facts and figures gathered in the survey. Colin Clark was there, but he was most amiable and did not put forward any of his theories. It sometimes does not occur to professors in universities that resources are not what are potentially available but what are actually available at any point in time and that distribution cannot be done mathematically but requires social and economic changes which take time and for which a Campaign such as the FFHC was necessary.

Other FAO Studies on Economic Development of the Third World

APART FROM the Food Survey we also began other studies to put the work of the Organization as a whole on the right course. The studies covered an extensive range of operations: what was needed to be done to combat hunger; what were the technical and economic possibilities of increasing production and improving distribution; how the knowledge then available could best be utilised. These were the kinds of questions that needed to be answered in an authoritative way and which would undoubtedly help the worldwide effort to awaken interest in the various approaches to the solution of the problems of hunger and poverty. These studies, twenty-three in number,* were to be prepared individually and cooperatively by both ourselves and the other International Agencies and would be supplemented by national studies.

All these studies were essentially interconnected and dealt with different aspects of the same central theme — the problem of economic development of the underdeveloped countries. The studies showed that although raising the levels of productivity and income in agriculture must provide the basic orientation to our activities, such an aim could not be effectively pursued in isolation, since it was intimately bound up with the dynamics of general economic growth and with the social milieu in which various institutional factors operated to enhance or retard that growth. The problem acquired new dimensions as the result of an unprecedented rate of population growth. And the farmer, upon whose prosperity and sense of security the well-being of the community as a whole largely rested, had emerged as the central figure in any scheme of economic and social betterment.

FAO's *State of Food and Agriculture* that year (1962) had two special chapters, one dealing with the levels of income among farm people in countries at different stages of economic development, and the other with the more practical problems of agricultural development in the less-developed countries. The study showed that farm income per head and per family was lower in almost all the developing countries than in other occupations; that productivity per man depended not entirely on differences in technical knowledge and equipment but also upon facilities for marketing, stability of prices and the system of land tenure; that where the percentage of non-farm population was small, this by itself placed a limit on the internal market and thus created a ceiling on the extent to which agricultural productivity could be raised. It brought out that agricultural development must keep pace with progress in other sectors while these sectors benefitted from a parallel development in agriculture. In the early stages agriculture must be the main source of

*See Appendix II on Page 308.

both manpower and investment resources. Agriculture was nonetheless a key sector in its own right.

The Mediterranean Development Project, which I will examine more fully in Chapter 18, demonstrated very clearly that unless action in agriculture and forestry was closely coordinated with the country's general economic development, it might accentuate imbalances in the economy. The broad conclusion was that the sectors which were most likely to respond positively should receive priority in spearhead action. The selection of spearhead sectors would have to be based on a comprehensive overall analysis, not merely of the economic conditions but also the social conditions in the country. This selective approach was based on the promise that efforts concentrated on a few favoured areas could be extended gradually to other sectors and supported by the increase in income, investment resources and taxable capacity of these "propulsive" regions.

The Consultant on Social Welfare brought in at that time was an eminent social scientist from India. When I had first proposed this study, it was argued by some countries that there could be no concrete measurement of welfare and that no criterion could be laid down for assessing the welfare content of FAO's activities. The report strikingly brought into focus FAO's overall role in the wide perspective of human want and international declarations for its amelioration. This fresh presentation of FAO's role served a useful purpose. It placed great emphasis on agrarian reform as "a basic prerequisite for effective work towards rural welfare" and suggested that FAO should give more attention to this subject than it had done in the past. On the basis of this report FAO's organizational structure was reviewed and a new Division was created to ensure a more balanced treatment of the different elements bearing on rural welfare.

The Forward Appraisal of FAO Programmes undertaken at the same time for the following five years, apart from streamlining and rationalising its activities, led to the broad conclusion that the time had come for a shift of emphasis towards an intensification of FAO's activities in agricultural infrastructure. Until then, FAO had concentrated attention on those major fields which involved a direct attack on low productivity. As a result, distribution and institutional and administrative problems had tended to be relegated to the background. Moreover, there had been little recognition of the fact that, because of the operation of market forces and other economic factors, the economic and social welfare of the rural population depended largely on the buying power of others.

Our thinking on all these matters was influenced by the Freedom from Hunger Campaign, as the content and direction of the FFHC itself were influenced by what emerged from these independent studies.

Chapter 13

The Freedom from Hunger Campaign

World Reactions

NEVER BEFORE has any World Campaign so permeated society at all levels in such a short time. The main reason is that hunger has a unique universal appeal. Nothing touches the conscience of man as much as hunger. It brings into man's immediate consciousness the social injustices and inequalities, the divisions between man and man that encrust social structures everywhere. The response to the Campaign was immediate, almost dramatic. Member Governments of both developed and developing countries, world leaders, international and national NGOs, and the intellectual community in every country expressed their enthusiasm. Pope John gave the Campaign his blessing and lent great prestige to its promotion. The Christian Churches of all denominations came forward with their worldwide moral and material resources to give it meaning. One Church leader from France said that the Campaign was proving a powerful ecumenical force and this was said against the background of the Ecumenical Council then in session in the Vatican.

Some of the basic concepts of the Campaign proved their value from the start. One of them was that National FFHC Committees were to provide cooperation between government and non-government representatives. The FAO Resolution had specifically asked for involvement of individual and private organizations and citizen groups. This strengthened the Campaign more than any other factor.

President Kennedy of the USA had from the very beginning shown great personal interest. He readily accepted my invitation to inaugurate the World Food Congress. I had chosen Washington for this first Congress because it was President Roosevelt who had called the Hot Springs Conference in 1943 in the middle of the war to bring FAO into existence and USA had given all the support necessary to get it going. Kennedy's address at the Congress was inspiring and showed his commitment better than any words I can use.

The President who made the deepest impression on me was de Gaulle. He was a towering personality, both physically and morally. His personal interest in the Campaign was sometimes a source of embarrassment to staff at Quai d'Orsay. I saw him several times. On one occasion, when I landed at Orly Airport it was surrounded by tanks. I was told that Paris was expecting the rebel French Air Force in Algeria to bomb the airport. When I went to the Elysée Palace I was told by the Aide that the President had only fifteen minutes to spare. The President talked to me in French through an interpreter but listened to me in English. There was no trace of anxiety or preoccupation on his face. He talked for nearly 45 minutes asking detailed questions about the Campaign.

I got a further insight into his mind in the message delivered by the French Minister of Agriculture at the FAO Conference in 1961:

> In our time, the only battle worth fighting is that for mankind. It is mankind we must save, serve and help to improve. We have enough food, enough clothes . . . we have mines and factories working at full pitch; fields which are well cultivated . . . all our children learn to read. We build numerous universities and laboratories. We train squads of technicians . . . we have enough doctors, hospitals and medicine to alleviate suffering, to take care of the sick and to save the life of nearly all the newborn children. Why do we not unite to build up the fraternal organization which will assist others? Why do we not pool a percentage of our raw materials, of our manufactured goods, of our foodstuffs? Why do we not make available part of our scientific, technical and economic staff, some of our ships and planes to eliminate poverty, develop resources and assist the less developed peoples in their daily toil? Let us do so but not to make them pawns for our politics, but in order to promote the chances of survival and peace.

Under President de Gaulle's influence, France was one of the countries where generous measures were used to spread the message of the Campaign. One of his chief aides in doing this was Maurice Schumann, then head of the Gaullist party in the Parliament.

Another person whose interest was most helpful was Prime Minister Harold Macmillan of Great Britain. When I visited him at No. 10 Downing Street he was about to chair a Cabinet meeting and several of his Ministers were waiting. Yet he took me to his Cabinet room to talk about the Campaign and then to the garden, where he showed me the tree where great Gladstone used to sit to compose his long speeches.

Britain has always been in the lead in humanitarian activities, especially through private organizations. These non-governmental organizations continued to be in the forefront all during the Campaign. Queen Elizabeth, in her Christmas message to the Commonwealth in 1963, made the FFHC her main theme: "Humanity can only make progress by determination and concerted effort. One such concerted effort has been the Campaign to free the world from hunger."

The President of my own country, Dr. Radhakrishnan, a long-time friend, used to receive me regularly and open his mind and heart to me freely. Here was another man the like of whom I shall never see again. He was a man of profound learning and a great philosopher. His interests were always concentrated on the conditions of the world's poor.

President Nyerere is well-known for what he has done for his people. I first met him in Rome when he called to see me. I was ill in bed yet he visited me nevertheless. His country was then on the verge of independence. He said he would rather wait till the question of federation of the three countries — Kenya, Uganda and Tanzania — had been settled than have independence for Tanzania on its own. A man of great idealism and honesty, I have always valued his friendship. In 1963 I invited him to deliver the McDougall Lecture. His Minister told me that for a few days he was in seclusion composing his thoughts for the occasion. His address stands out as one of the best in the series.

President Lübke of Germany first came to the FAO Conference in 1959 as Minister of Agriculture and we became close friends. On his invitation, I visited Germany several times going twice to West Berlin for the Green Week and once on a tour across the country. He was a very warm-hearted person and a source of great encouragement to me.

Presidents Marcos and Pak Chung Hee of South Korea and King Bhumibol of Thailand were among others whose genuine interest in the Campaign must be put on record.

I travelled widely to keep in touch with the member countries and leaders of public opinion working for the FFHC. I travelled through tropical rain forests, the blazing deserts of the sub-Sahara, in the ice and floating icebergs of Greenland. One Arab country found me unacceptable. The incident arose over a very minor matter. An English officer had been transferred to that country to work as a fruit-packing expert. His luggage contained a package marked "Made In Israel". It was seized and the officer was to be arrested when he arrived from Rome. Under the Protocol signed by the member countries, UN officers enjoy diplomatic status and are exempt from criminal proceedings of this kind. I asked the officer not to proceed and to get his luggage or whatever remained of it back to Rome. This incident apparently went on record against FAO and when I expressed a wish to visit that country some time later various excuses were put forward which made the visit impossible. I did not visit South Africa for obvious reasons though their delegate had expressly invited me.

Chapter 14

Progress of the Campaign

IN RELATING the progress of the Campaign culminating in the World Food Congress in Washington DC in June 1963, the best course would be to reproduce here relevant extracts from the four consecutive letters I addressed to the Ministers of Agriculture of our Member Countries during this period. The letters convey a sense of urgency and give a "feel" for the atmosphere created by the Campaign at that time which cannot quite be recaptured in retrospect after the lapse of so many years.

Extracts from Letters to the Ministers of Agriculture

Since the Freedom from Hunger Campaign was officially launched on 1 July 1960, two very significant developments have taken place within the United Nations family — the Development Decade and the World Food Programme. Like the world Campaign against hunger, they spring from a deepening concern with the economic and social imbalances in the world today and reflect the hopeful vision of a future of economic freedom and social justice for all.

The problem of hunger and malnutrition has always been with man and has made man's history. But what has invested it with a new dimension and charged it with a new urgency is the explosive rate of population growth with the advance of science and social hygiene. The Freedom from Hunger Campaign is intended to be primarily educational in character — to make the Governments and peoples all over the world aware of the nature of the problem so that integrated efforts can be made both nationally and internationally to overcome it. Fund-raising is not the primary objective of the Campaign, although action projects have an important part to play by providing a means through which developed countries can express their solidarity with the developing countries, assisting their efforts to implement their national development plans, and making available an additional source of skills, technical know-how and foreign exchange which in a developing economy can make all the difference.

It is recognised that no lasting solution to this problem can be found without balanced economic and social development. The emphasis of the Campaign has therefore been on the transforming of subsistence agriculture into a market economy, on increasing the productivity of land, water and forestry resources, and on quantitative and qualitative increase in the production of food. Hidden hunger, or malnutrition, which is the parent of many diseases prevalent in large segments of the world's population can be cured only with provision of more plentiful supplies of food. Besides, agricultural development has to be the spearhead of economic and social development in the developing countries where the bulk of the population lives on the land. The essential savings required for investment in development programmes can accrue only if the income levels of rural communities are enhanced through increased productivity. There are many problems to be solved before the vicious circle of rural poverty can be broken, but a start has to be made on two major objectives — greater production of food and higher purchasing power of the rural masses in the developing countries.

The relationship between the objectives of the Freedom from Hunger Campaign and those of the United Nations Development Decade is an intimate one. The basic aims and approach are the same, though there is a difference in emphasis and in method.

The General Assembly Resolution (7710 XVI) which designated the sixties as the UN Development Decade was adopted on 19 December 1961 (a little over two years after the Freedom from Hunger Campaign Resolution was adopted by the FAO Conference). The Resolution calls the Development Decade a programme for international economic cooperation. While it asks the developing countries to intensify their own efforts toward self-sustaining economic growth, with each country setting its own target, taking as the objective a minimum annual rate of growth of aggregate national income of 5 per cent at the end of the Decade, it also urges the developed countries to provide sustained support for accelerated growth in the underdeveloped countries, specifying the broad measures which they should follow for this purpose.

The references to food and agriculture in the Resolution are particularly significant in view of the emphasis placed in the debate by some speakers on industrialisation as the major avenue for rapid economic growth. In both my speeches at the ECOSOC session (1962), I argued that there could be no successful industrial development without rural development and that the two mutually reinforced each other. I particularly warned against the grave risks which countries with predominantly agricultural economies ran in attempting industrialisation in a great leap forward and neglecting agriculture in the process. Agricultural progress must supply the food for better nutrition of an increasing population; agricultural progress must release the labour for industrial development; and agricultural progress must supply in terms of export returns and internal revenues much of the foreign exchange and capital to finance imports and capital investment. This was the role which agriculture played in the economic development of the most advanced industrial country in the world today, the United States. This is the role which agriculture must play in most of the developing countries in this decade and in the decades immediately ahead.

The difference in method lies in the fact that the Campaign not only relies on the support of Member Governments but also invokes public participation and involvement of citizen groups in its world objectives and action programmes. This emphasis on public participation is a distinctive feature of the Campaign and has stimulated mass support to the extent that we now see the Campaign developing gradually into a concerted world movement.

The World Food Programme, set up earlier this year in terms of the UN General Assembly Resolution 1496 (XV) of 27 October 1960* is an international response to the stimulus of the Campaign and is of particular significance. The text of this Resolution indicates the relationship between the World Food Programme and the Freedom from Hunger Campaign. The World Food Programme necessarily involves a heightened activity for FAO. Even so, it only covers a segment of FAO's much wider responsibilities. The objectives of the Campaign and those of the Charter cannot be achieved by the use of "surplus food" alone

*See Appendix III on Page 309.

which is the instrument of the World Food Programme. The Programme only seeks to provide one important element amongst the many required to contribute to the success of the Campaign, and hence to the fulfilment of the ideals of this Organization.

The Development Decade, the Freedom from Hunger Campaign and the World Food Programme — each has its own significance in an inter-related whole. The Development Decade is an effort to encourage all the activities which must contribute to economic development. The Freedom from Hunger Campaign is dedicated to the solution of the central problem not only of this Decade but of the many decades to come — the need to increase agricultural productivity in the developing countries, where the vast majority of the people derive their livelihood from agriculture. The World Food Programme is specifically directed to the use of food as one means of achieving the wider objectives of the Campaign and the Development Decade. While food aid through the World Food Programme could be used to support broad projects conceived under the Freedom from Hunger Campaign, the Campaign could in turn be used to provide non-food support for projects under the World Food Programme.

Special Assembly on Man's Right to Freedom from Hunger (March 1963)

IN SUPPORT OF and in preparation for the World Food Congress, to be held in Washington in June 1963, I undertook two actions of importance. I invited some leaders of contemporary world thought including several Nobel Prize winners to come to Rome on 14 March with the object of bringing their moral authority to bear on the aims and purposes of the Campaign launched by FAO. This March Assembly was also intended to provide a dramatic opening for the World Freedom from Hunger Week which followed immediately after.

The Special Assembly on Man's Right to Freedom from Hunger was held in Rome on 14 March. The Assembly was attended by twenty-eight world-renowned personalities and others whose contribution to the thought and culture of the twentieth century or whose concern for the problem of Hunger was well known. Not since the Campaign was launched had such a range of interest been displayed in the problem of world hunger, nor had such a forceful analysis of the problem been presented from a single platform. The pronouncements of this eminent group of men and women revealed a penetrating awareness of the social and political implications of many hundreds of millions of people being condemned to a life of hunger and want. They also put on record the need to mobilise the attention and the action of people everywhere if a courageous and rational improvement of this human condition was to be effected.

The Assembly's Manifesto* highlights with singular clarity some basic questions of fact and policy which the World Food Congress could not afford to ignore if it was to fulfil its purpose. Can food production keep pace with population growth unless all available human and material resources are utilised for production purposes within a framework of rational planning? Enough scientific knowledge and technological experience were available for bringing about an agricultural revolution in the poorer countries, but could this knowledge be applied within a social and institutional framework that had held up progress for centuries through its own inertia? Could external aid by itself be effective in stimulating economic growth in the absence of world commodity agreements which guarantee fair and stable prices for the products of developing countries? Could the current inadequate levels of investment in development be significantly increased without drastically cutting down through international agreements the present astronomical scale of expenditure on armaments?

"Freedom from hunger," states the Manifesto, "is man's first fundamental right. In order to achieve this, we suggest urgent and adequate national and international effort in which the Governments and the peoples are associated. More particularly, we desire to draw attention to the colossal waste of resources in the piling up of more and new forms of armaments and the immense assistance to the Campaign against Hunger that even a partial diversion of these funds could achieve. We feel that international action for abolishing hunger will reduce tension and improve human relationships by bringing out the best instead of the worst in man."

Freedom from Hunger Week (17-23 March 1963)

THE MAIN OBJECT in observing the Freedom from Hunger Week was to heighten the feeling of world solidarity in winning man's first freedom — freedom from hunger. Falling as it did at the mid-point of the Freedom from Hunger Campaign, the Week was an invitation to individuals and groups to participate in a tangible way in furthering the objectives of the Campaign. Addressing letters to some 120 Member Governments of the United Nations family, I suggested a few lines of action which I considered to be appropriate for the observance of the Week. These suggestions included the issuing of proclamations by Governments anc

*See Appendix IV on Page 314.

messages by Heads of States, strengthening of the National FFHC Committees, organization of agricultural and nutritional development programmes, adoption of laws and administrative measures relevant to the Campaign, religious observances, introduction of teaching programmes on FFHC into national educational systems, special coverage of Campaign and Week activities by mass communications media and participation by citizen and professional groups in events specially organized for the Week. These by no means exhausted the possibilities of giving the Week a truly national flavour while keeping in view the larger international world objectives. In fact, some of the reports I received regarding the planning of national observances indicated a variety of imaginative programmes were being adopted in various parts of the world for the occasion.

I might mention here a few examples of the specific action that had been taken by some of our member countries. President Kennedy had issued a Proclamation urging "the American Freedom from Hunger Foundation to take national leadership in planning appropriate observance of this Week, and American citizens in all walks of life to participate in the observance of the National Freedom from Hunger Week".

President de Gaulle's interest in the whole range of problems centred on hunger found expression in a personal message which was made available to us for Campaign promotion and was widely broadcast. Following this message an elaborate programme drawn up by the French National Freedom from Hunger Committee enlisting the services of press, radio and television was initiated. The Canadian Freedom from Hunger Committee decided to hold a National Food Conference from 19 to 21 March at Ottawa. The Freedom from Hunger Campaign Committee of the United Kingdom planned a most comprehensive and imaginative programme on the theme "An Hour for the Hungry" which involved practically all sections of the population. I should like to draw special attention to the French and UK programmes as they contain practical models for the developed countries. A special feature of the observance in Ireland was fund-raising in primary schools. The student community of the Netherlands collected funds during their holidays and Lebanon initiated a nation-wide educational and fund-raising campaign during the Week.

Most of these activities were undertaken by the National FFHC Committees or NGOs and other citizen groups. This was as it should have been, since the main purpose of the Week was to secure public involvement and commitment in the objectives of the Campaign. The observance of the Week was not necessarily limited to the period 17 to 23 March, although the central date of 21 March was chosen because nearly 150 national administrations had decided to issue special FFHC stamps on and around that date. In fact, some of the activities were to continue well after 23 March.

The United Nations and several of our sister Agencies came forward with declarations and projects to emphasise their solidarity with FAO. The Secretary-General of the United Nations and the Executive Director of UNICEF issued special messages stressing the importance of the Week. The World Health Organization designated "Hunger: Disease of Millions" as the theme of its annual Health Day which was held that year on 7 April. UNESCO had asked its National Commissions to participate in the Week through appropriate programmes organized by the International Council of Museums, the International Music Council, the International Theatre Institute and similar cultural bodies. The Human Rights Commission of the United Nations had been approached to associate the fifteenth anniversary of the Universal Declaration of Human Rights with the Freedom from Hunger Campaign in a suitable manner. Other specialised agencies planned appropriate action. These activities would also generate interest in those sectors of the population which might not be covered by the National FFHC Committees.

Several of the international NGOs, who had already made significant contributions to the development of the FFHC, readily lent their support to the Week in a positive and concrete manner. These NGOs had a combined membership running into millions all over the world and their efforts would add considerably to the educational impact of the Week. I should like especially to mention here an event of outstanding importance that had been planned by the World Council of Churches. The Council held a special service on Thursday 21 March in the Cathedral Church of St Pierre in Geneva, and requested all member churches, covering 350 million people in 80 countries, to hold similar services on Sunday 24 March. Other major world faiths also observed the Week in appropriate ways.

While there was enthusiastic support for the Week in every country, some of our Member Nations found themselves too preoccupied with their day-to-day problems to organize a programme. And yet it was difficult to see how the tasks of national development could be accomplished without solving the basic problem of providing food for the people. I suggested to them that the Week could be observed in several ways without taxing their resources unduly. First of all, the Week might serve as the occasion for stimulating public discussion of the national development plans and the role assigned to agricultural development in them. Secondly, the Week might provide an opportunity for adopting nutritional programmes with broad-based public participation in them. Thirdly, the Week might be utilised for intensifying the programmes of community development where they existed, and initiating self-help programmes in the villages aimed at better nutrition, health, cleanliness and rural education simultaneously with improved farming practices.

It seemed to me that such a programme could be effectively put into practice only if local leadership in the rural communities could be mobilised for the purpose. It was the village or community leader who enjoyed the confidence of the rural people and could command their participation in a cooperative effort for the improvement of their living conditions. There were many improvements that could be achieved without excessive capital outlay. Unemployed or under-employed people might be induced to contribute their labour for the improvement of community life. Village fairs in the developing countries are often an excellent medium of communication for the rural people. Special displays and exhibits could be produced at low cost. Film shows could also be arranged at community centres. The Week might provide the occasion for initiating rural broadcasting programmes. Tree-planting may also be suitable in areas which were not too arid in March. There were many ways in which the message of the Week could be brought home to the rural masses of predominantly agricultural countries. The programmes naturally would have to take account of local needs and local conditions.

The Week was symbolic in another sense. March 21 was the spring equinox in the northern hemisphere and the autumn equinox in the south. At this time, men's minds turn to sowing and harvesting. Many religious festivals that take place during this season have their roots in ancient traditions. It is a time for prayer and for work.

The World Food Congress (June 1963)

IN JUNE 1963, we were approaching a crucial phase in the development of the Freedom from Hunger Campaign. During the past three years an intensive public education campaign had been carried out with satisfying results. The realities of the current situation regarding the extent of hunger and malnutrition in the world had been brought to the notice of millions of people through the efforts of governments, citizen groups and communications media. The Freedom from Hunger Stamp issue, in which some 150 postal administrations had cooperated, had brought the message of the Campaign to the notice of an estimated 1,000 million people throughout the world. Public opinion, however, had not been content with knowing the facts alone; it also demanded leadership in cooperative action on a world scale to solve the problem. This was a complex matter, and the facts of economic and social life in the developing countries was such that no panacea could be prescribed. These facts had to be analysed and sifted and all existing knowledge brought to bear on the formulation of suitable action for each individual situation. The World Food Congress was held not only for the purpose of bringing the various approaches under a single focus, but also to provide a lead by indicating possible lines of action.

The World Food Congress brought together thinkers, scientists, sociologists and representatives of citizen groups from all parts of the world who conferred together on the challenge of hunger that confronts mankind and were ready to give a lead to world opinion on what should be done to meet this enormous challenge.

The Congress was held in Washington. The atmosphere was one of urgency. The tone of the Congress was set by President Kennedy in his inaugural address, and by President Radhakrishnan of India who was in Washington at the time on a State visit. The young President of one of the most advanced countries spoke with earnest intensity and declared that failure to conquer world hunger would be "a disgrace for this generation". The slight, white-haired fragile-looking President of the country that reflected all the main problems of world hunger stated his faith, "this Congress will arouse the conscience of the peoples of the world". From this moment onward, the discussions in the Congress were carried out in an atmosphere of high purpose and shared responsibility, and at the concluding session a Declaration was unanimously adopted pledging the world to a continuation of the struggle against hunger by means of all the resources, physical as well as spiritual, available to mankind.*

As the main objective was to focus world attention on the basic aims of the Freedom from Hunger Campaign, the Congress was deliberately planned as a "people to people" meeting, the participants attending in their personal capacity and having full freedom to express their views critically. It was realized that such a Congress could not be asked to formulate a detailed "master" plan of action or to come to concrete decisions to be carried out by Governments. But it could generate an impulse to action which would be geared to the urgency of the situation.

The interest that the Congress created was evidenced by the presence of more than 250 newspaper correspondents, editors and radio and TV commentators. Publicity arrangements included worldwide coverage by the main news agencies. On the whole coverage in leading newspapers of the world and over radio networks was excellent, particularly in America. Not only did the World Food Congress appear constantly in such influential papers as the *Washington Post* and the *New York Times*, but the news and feature stories were frequently supported by editorial comment. Highlights of the coverage of the Congress were President Kennedy's opening speech, my National Press Club speech and the speeches of Messrs. Toynbee, Myrdal and Palewski; but I think it is fair to say that all the main speakers were given good coverage, particularly on a regional basis. Reports that came in indicated that coverage was good in all parts of the world.

The attendance at the Congress exceeded the original expectations. I had planned on the basis of an attendance of 1,200. The actual registration exceeded 1,300. What was more important, the participation from the developing countries was an encouraging feature.

*See Appendix V on Page 316.

It would be invidious to single out individual contributions, but some need to be mentioned. U Thant, Secretary-General of the United Nations, said that the problem of hunger and malnutrition required the cooperative effort of all the UN family and that FAO was making a most important contribution to the Development Decade of the United Nations. The Heads of the other UN Agencies who spoke after him pledged their continued support for the Campaign.

Of the other distinguished speakers, Dr. Toynbee referred to the threat posed by unregulated human reproduction, and the need to persuade ourselves to give to the interests of the human race as a whole a decisive priority over the interests of one's own particular section. Monsieur Palewski spoke about the paradox resulting from the co-existence of overproduction and undernourishment; how excess man-power could be turned into an asset instead of being a liability; and made a stirring plea that savings resulting from a real disarmament should be utilised for the development of underprivileged regions. Professor Myrdal said that no drive for industrialisation had a chance to succeed if not preceded and accompanied by a decisive rise in agricultural productivity. He also said that if we assumed that international aid should be based on international solidarity, there was no reason why the burden of aid should not be carried by all rich countries, whether they had food surpluses or not, according to some principle of equitable distribution related to their income. Professor Bovet touched on the future prospects for new sources of food. Mr. Krishnamachari gave the lessons learned from eleven years of planning in India. Among the interesting proposals which came out were the following:

(1) The creation of an International Training Institute to train personnel for investigation and introduction of progressive land use systems, including resource assessment and land use planning on the physical side, combined with a regional study of the economic and social implications of agricultural planning as an integral part of economic development plans.

(2) The formation of a "pool" of food production requisites under FAO supervision on which developing countries could draw as need arose, associated with the World Food Programme.

(3) The establishment of spearhead development zones, such as FAO Mediterranean Development Project, to serve as focal points for the speed of agricultural and general economic development.

(4) Increase in the scale of external assistance to developing countries by the creation of an "International Solidarity and Development Fund", mainly through reduction in the steadily increasing military expenditures.

(5) In order to ensure that the world might be freed from hunger within the foreseeable future, the formulation of a world plan, in quantitative terms, on the basis of nutritional needs, indicating the type and magnitude of external assistance needed in relation to

local resources, and internationally coordinated.

(6) The creation, through FAO, of an international liaison unit with Western Food Industries, to harness the resources of this industry for assisting through FAO the new food industries in the developing nations.

(7) Initiation of action through FAO and other suitable international organizations to provide assistance in establishing or strengthening applied research facilities, preferably on a regional basis.

(8) The creation of national nutritional institutions for training, preparation of food-balance sheets, nutritional education, food-management and for influencing the direction of farm production.

(9) Measures to provide for cooperation among the non-governmental organizations, between non-governmental organizations and National FFHC Committees, and between governments and international agencies.

(10) Placing of the Freedom from Hunger Campaign and the National Freedom from Hunger Campaign Committees on a continuing basis and widening and strengthening the FAO coordination of FFHC work.

A recommendation of great importance was for the periodic holding of a food congress, at which the Director-General of FAO would present a world review of the food situation in relation to growth of populations in all the developing countries, together with a proposed programme of future action, taking into account programmes of national development and institutional reforms in various countries, the findings of technical and economic services and the scope and problems of international cooperation. In this venture the Director-General would have the assistance of high-level committees appointed for this purpose in each country participating in the Freedom from Hunger Campaign.

The Congress was held on the twentieth anniversary of the Hot Springs Conference. The question that naturally arose in many minds was whether we had reason to be optimistic when so little progress had been made during the previous two decades. There were several reasons to make us hopeful. Firstly, we had a greater certainty of possessing the know-how. During the recent past, countries like Japan and Mexico had shown that production could be increased many-fold within a short space of time. Secondly, planning at the national level and coordination at the international level were accepted as operational imperatives. Thirdly, the urge to mobilise internal resources with a view to accelerate progress had become pronounced in all the newly independent countries. Fourthly, a new concept of international solidarity to promote global well-being through external assistance had emerged, and "foreign-aid" was understood more and more as "investment in mutual well-being".

The possibility of overall and rapid action at all levels was thus greater than ever before. The declaration that was passed unanimously and with acclamation on the last day of the World Food Congress was a true expression of this new spirit and new determination. The general political situation also afforded cause for hope. The possibility of a nuclear test ban might prove to be a definite step forward in world disarmament. The implications of a test ban and disarmament in the fight against world hunger was made a central theme at the March Assembly which met in Rome early that year. The celebration of the fifteenth anniversary of Human Rights was due. Up to then attention had been generally concentrated on civil rights rather than on freedom from want. It was suggested that celebration of the Anniversary could be made more meaningful if nations determined to give more attention to freedom from want in their plans and programmes in future.

At the FAO Conference later that year I gave a *resumé* of the progress of our Campaign and the results of the World Food Congress; I then referred to the address of Christopher Soames, UK Minister of Agriculture who had drawn attention to the UN resolution on Hunger, Disease and Ignorance and had invited all NGOs, National FFHC Committees, Church groups and others to throw their weight behind this wider Campaign. I pointed out that only a few days previously the National FFHC Committees and non-governmental organizations had gathered together and there had been a call for the organizing of all the National Committees on a permanent basis, and a suggestion that a World Congress be held periodically to review a world survey that was to be prepared by the Director-General dealing with the world food situation in relation to population and overall development and also to review a proposed programme for future action. After I made these points the delegates expressed themselves against the idea of a wider Campaign.

I added that whether the Campaign was carried on in its present form or with a widened mandate to give more emphasis to health and education, there was consensus that the essential factor for success was involvement of the people, supported by governments, and guided and coordinated by the UN Organizations concerned. The proposal to include the problems of illiteracy and health was a tribute to the work that the National Campaign Committees and NGOs had done. However I had serious doubts about the suggestion that the very success of the FFHC should be the reason for dismantling it and replacing it with a bigger Campaign under another name.

In Chapter 16, I will deal more fully with the debate in ECOSOC on the General Assembly Resolution for a wider Campaign. I will only say here that the 1963 FAO Conference decided to continue the Freedom from Hunger Campaign until 1970.

I may interpose some personal reminiscences here to complete the scene of the World Food Congress in Washington. A few days before it was due to start — it was being held in the State Department Building —

I was approached by a very senior officer with the proposal that Secretary-General U Thant and not the Director-General of FAO should open the Congress alongside President Kennedy. The argument was that U Thant was the head of the UN system and in this World Food Congress the whole UN system was participating. I pointed out that the Congress was being held as an integral part of the FFHC which was being directed by the FAO Director-General according to the FAO Conference Resolution. I had the highest respect for U Thant both personally and as Secretary-General but this was not a matter of courtesy but of principle. I said that if the host country insisted I would have no alternative but to wind-up the operations at that stage and return to Rome. The proposal was dropped. I would like to say here that apart from the question of principle, my objection also came from my adherence to the idea that at any cost the strength of FAO as a non-political forum for the world to discuss problems of hunger must be protected.

The opening days of the Congress were overshadowed by two sad events. The death of Pope John removed from us a man who loved ordinary people and understood their problems. His death was a great blow to me for, as I will soon relate, I had known him very well. The other sad event at that time was that Herbert Hoover was lying on his death-bed. On behalf of all those assembled from the four corners of the world, I sent a message with our prayers for his recovery and our homage to his great humanism and compassion for stricken people everywhere.

I may recall here my brief but close association with Herbert Hoover during his visit to India in 1946 and the deep impression he made on me. He had invited me to come and see him if ever I should be in the United States. Six months later, I was in Washington. I visited him at his apartment in the Waldorf Astoria Towers in New York. He had just returned from an official mission to Germany to dismantle all industrial complexes which could be put to use for war in the future. He confessed to me his frustration at such a mission, at a time when all western Europe, and Germany in particular, needed rehabilitation and reconstruction. The portents of the "Cold War" were already visible.

The World Food Congress was a landmark in FAO's history in every sense. The Congress ended with a note of discord though fortunately it was a behind-the-scenes affair. I had prepared the Declaration to be placed before the Congress for adoption only two or three days before the case. I had no time to consult with the Chairman, the US Secretary of Agriculture. When I showed it to him, he appeared disturbed. At a dinner party that same evening he was sitting next to me and told me that he was opposed to the terms of the Declaration since it went too far to the left and would refuse to preside over the session if I pressed it. I could not convince him that there was really nothing to which anyone, knowing the prevailing political climate in the world, could object. He refused to preside and I took over. I asked the assembled delegates to stand as I read the Declaration and it was adopted by acclamation.

Young World Assembly
(October 1965)

FROM THE BEGINNING a thought shared by all was how to engage the Youth in a meaningful way in the FFHC. In October 1965, I called together youth leaders from all parts of the world for a dialogue. The Manifesto issued by The Young World Assembly has an appeal all of its own, and I reproduce it in Appendix VI.* The FAO Conference (1965) had this to say:

> The Conference warmly welcomed the launching of the Young World Mobilisation Appeal, and agreed that it could be of great value for the pursuit of the objectives of FFHC. Several delegations gave examples of the mobilisation of young people for local action such as the construction of schools, training centres, small irrigation schemes and roads. One delegation explained that the work being done by young people had proved to be extremely valuable in helping the country toward self-sufficiency in its staple crop. Strong stress was laid on the importance of ensuring that programmes under the Young World Mobilisation Appeal should be designed to provide opportunities of constructive action for young people, and not merely to enlarge youth organizations.
>
> The recommendation of the Second FFHC Conference, that a Director-General's Promotion Group for the Young World Appeal should be established within the framework of FFHC, was welcomed by the Conference and it was stressed that there was an urgent need for a planned programme to follow up the initiatives already undertaken both by the Director-General as well as by countries and organizations.

I kept up the momentum of the appeal of the Young World Assembly in various ways. My last direct contact with the Youth representatives was in Toronto in 1977. I addressed a Young World Food and Development Seminar, where I met the rising young African Leader, Tom Mboya of Kenya, who died a few years later at the hands of an assassin, a great loss not only to Africa but to the world.

*See Page 320.

Chapter 15

The Vatican

My Relations with Pope John

IT WAS MY good fortune to enjoy the confidence and goodwill of both Pope John and Pope Paul. It was the Freedom from Hunger campaign that brought this about. On the inauguration of the Freedom from Hunger Campaign in July 1960, I had the following message from Pope John:

> On the occasion of the official launching of the Freedom from Hunger Campaign organized by the very worthy Food and Agriculture Organization of the United Nations, We wish to renew Our most paternal encouragement for this very generous initiative that corresponds so well to the true welfare of mankind and deserves so much to enlist the interest and collaboration of all men of good heart.
>
> The Church — as We have already had occasion to state — rejoices to see so many men of goodwill uniting for the successful accomplishment of this great undertaking. The Church is happy to note that the Campaign is strikingly designed to promote those good works which It so warmly recommends that Its children practise.
>
> We, for Our part, would accompany with Our supplications and Our prayers the efforts of all those persons and institutions that will be taking part in the Freedom from Hunger Campaign, and with all Our heart We pray that Heaven will abundantly bless them.

On numerous occasions after that, I asked for the Holy Father's audience for Delegates gathering from different segments of society which he readily granted and blessed with words of warm encouragement. In his famous Encyclical *Mater et Magistra,* he mentioned FAO in these words:

> It is evident that both the solidarity of the human race and the sense of brotherhood which accords with Christian principles require that some peoples lend others energetic help in many ways. Not merely would this result in a freer movement of goods, of capital, and of men, but it also would lessen imbalances between nations . . .
>
> Here, however, We cannot fail to express Our approval of the efforts of the Institute known as FAO which concerns itself with the feeding of peoples and the improvement of agriculture. This Institute has the special goal of promoting mutual accord among peoples, of bringing it about that rural life is modernised in less developed nations, and finally, that help is brought to people experiencing food shortages.

The message he sent on the occasion of the World Freedom from Hunger Week in March 1963 showed the breadth and depth of his mind:

> In Our concern with being faithful to the doctrine of Christ and in accordance with the purest tradition of the Church, We have been pleased to encourage, since its inauguration in 1960, the Freedom from Hunger Campaign so laudably launched and pursued by the Food and Agriculture Organization of the United Nations.
>
> At a time when, within the framework of this vast Campaign, the World Freedom from Hunger Week is about to begin, and when the World Food Congress is approaching, We want to say how opportune are these new initiatives, in Our opinion, and how much We wish that they may benefit from universal collaboration.
>
> The problem is really to put human energies to work on a very great scale. In the minds of its originators, this noble undertaking has only one aim — which would of itself be meritorious and would command respect — to alleviate temporarily the deficiencies of peoples who are in the process of development. It aims primarily at provoking the unanimous effort, of all those in a position to make it, to teach mankind how to use fully the super-abundant gifts which the Creator has placed at the disposal of humanity.
>
> How many are the expressions of admiration, in the Holy Scripture, at the marvels of Creation and the bounty of God, who has bestowed upon man "dominion over the works of (His) hands" and "dost crown him with glory and honour" (Psalms viii, 5 and 6).
>
> This control of man over nature, in the words of the inspired author, is every day revealed to be more widespread. Modern

means of investigation provide a glimpse of the still almost-unexploited treasures hidden in the bowels of the earth and the depths of the oceans. It is man's duty to use the gifts of intelligence and will which he has received in striving to develop these immense riches.

But it is also the immediate duty of society, with the resources at its disposal, to bring concrete assistance to those of its members who are deprived of the minimum essential to the normal flowering of their personality. St. Paul's warning to the Galatians is still valid, and now more timely than ever: "Bear one another's burdens, and so fulfill the law of Christ." (Gal. vi, 2).

Considering the prodigious increase of transportation and travel facilities in the modern world, one can no longer say that the hunger and malnutrition prevalent in certain regions of the globe are due solely to the insufficiency of natural resources presently available, since those are super-abundant in other regions. What is missing is the organizing and coordinated intelligence and will capable of ensuring a fair distribution. Also lacking, among developing peoples, is the adequate exploitation of their own resources.

May this World Week for the fight against hunger — and the approaching World Food Congress in Washington — be a call and a stimulant to all men of goodwill. May they strive hard to speed up projects of agricultural development, to hasten — in line with the findings of the Geneva Conference — the application of science and technology for the benefit of the less-developed regions. In brief, may they strive to promote everywhere a better utilization and a better sharing out of human and material resources. In so doing, they will be sure to win the praise and gratitude of all righteous men, and to merit divine blessings in all their fullness. These We personally, wholeheartedly, invoke upon the organizers of these meritorious initiatives, and upon all those — persons and organizations — that will participate in them or benefit from them.

Pope John was indeed a great man, the likes of whom is given to mankind only at rare intervals in history. It is only natural that the Ecumenical movement should have come from him. I am sure I am not committing a grave act of indiscretion when I recall his telling me that when he first raised the idea of such a move he had nothing but opposition from those around him. But he set their objections aside.

The World Food Congress was held in Washington in June 1963. I knew that at the Congress the question of population control would figure prominently and if I could get a more positive message than what the Pope had said in the Encyclical it might give the world gathering a new dimension. After much thought, I addressed the following letter to him through the Secretary of State, Cardinal Cicogniani. I will quote the letter *in extenso*:

If I make bold to approach Your Holiness today it is because the question which I bring affects the future of mankind, and is one that must be faced. I have been deeply moved by Your Holiness's Encyclical Letter *Pacem in Terris*; the profound wisdom and understanding that emanates from it sheds a new light on the path that lies ahead of us. In the history of mankind, it is such inspired wisdom born in faith that has proved to be the greatest moving force in progress.

In your Encyclical Letter you declare that any human society, if it is to be well-ordered and productive, must lay down as a foundation the principle that every human being is a person; which means that his nature is endowed with intelligence and free will, and that, by virtue of this, he has "rights and duties of his own, flowing directly and simultaneously from his very nature, which are therefore universal, inviolable, and inalienable". The first of these "universal, inviolable, and inalienable rights" is defined as "the right to life and a worthy standard of living" comprising "the right to the means which are necessary and suitable for the proper development of life", including primarily food and other necessary goods and services, as well as "the right to security ... in any (other) case in which he is deprived of the means of subsistence through no fault of his own".

The first of the "universal, inviolable, and inalienable rights" is thus defined as a *dual* right, that is, not merely the right to life but also the right to a worthy standard of living. The text of the Encyclical does not itself comment on the implications of the stipulation of this dual right, affecting, as it does, the issue of the rapid rate of world population growth due to the advance of medical science and public hygiene. Yet I trust that in your profound wisdom you will already have given thought to these implications. The problem before mankind is that whatever practical new measures be taken nationally and internationally, now or in the near future, aiming at the accelerating expansion of productive capacity for the provision of supplies of food and other means of livelihood, a continuation of present trends of population growth may well lead to a situation when "the right to life" of those not yet living, that is, of larger and larger numbers of future generations, will no longer be a right parallel to, but will be in conflict with, "the right to a worthy standard of living".

The magnitude of the challenge facing mankind is brought to the fore by some of the data presented in a report I have just issued on behalf of the Food and Agriculture Organization of the United Nations, the *Third World Food Survey*. This Survey is based on food-balance-sheet data of over 80 countries, covering some 95 per cent of the world's population, and on food consumption and dietary surveys conducted in various parts of the world. The criteria of calorie and protein requirements used in this Survey, together with adjustments made for climate, age and work habits, are internationally accepted by nutritionists and supported by clinical evidence.

According to this Survey 10 to 15 per cent of the world population, that is 300 to 500 million people, actually go hungry, and up to one half of mankind suffer from either undernutrition or malnutrition, or both. By the end of this century, i.e. in another 37 years from now, food supplies will have to be increased fourfold in the Far East, threefold in the Near East, between two and threefold in Africa, and between three and fourfold in Latin America. The basis of our calculation is a diet which provides on the average 2,400 calories per person per day in the developing areas, and includes 70 grams of total protein of which 20 grams are of animal origin — a minimum average which, according to nutritional science, is essential to eliminate hunger and malnutrition from the world.

In another recent FAO publication, we have examined the physical possibilities of growing more food in different regions. We find that with proper investment in skills and capital and more intensive and extensive application of science and technology, supported by the necessary sociological changes and institutional measures, the problem is not insuperable in Latin America and Africa, though more difficult in the Near East with its vast desert areas and limited water supply. This is not to say that in these regions the problem can be left to take care of itself. But where the problem gives cause for the greatest anxiety is in the Far East, where half the world's population is concentrated, where, at the present trends of growth, the population by 2000 A.D. will exceed the present total world population. In a special study we have carried out in one area in the Far East, where environmental conditions are exceptionally favourable, we have found that, on our present knowledge of science and technology, a fourfold increase of production would be about the limit of achievement. The average potential increase over the whole of the Far East, with widely varying conditions, is certainly lower than in this favoured area and probably not more than threefold in any case. From these surveys one can well visualise serious shortages and even famines breaking out over increasingly wider areas 20 years hence, unless determined steps are taken to anticipate such developments.

It is against this background that I approach Your Holiness. We realize that there is no easy answer. We realize that we must take all possible measures to raise output and productivity per acre, to explore the resources of the land and of the seas, to find new sources of food by chemical and other processes. These were in fact the ideas that prompted me three years ago to call upon the FAO

> Member Nations to launch the Freedom from Hunger Campaign. Yet, while the objective of increased productivity and exploration of new resources must be ceaselessly pursued, the problem of population growth is a factor which needs to be forced. Otherwise the prospects for raising incomes per head in the poor areas of the world to levels which, in the words of the Encyclical, "will enable every citizen to live in conditions in keeping with his human dignity" are gravely jeopardised.
>
> Indeed, the world faces an unprecedently grave moral and physical crisis, though at the same time unprecedented opportunities for progress. Men and women of goodwill all the world over look to your wisdom for guidance.*

I left for Washington. On June 3, as I was addressing the Washington Press Club, the message came through that Pope John had passed away. I feel sure that if he had lived, the stark realities of hunger in the world would have moved him to give a more positive guidance to the world.

My Relations with Pope Paul

POPE PAUL in his earlier years followed the spirit of Pope John on some of these controversial questions. On numerous occasions he blessed FAO for its work for the hungry people of the world. On Sunday, 21 November 1965, in his address to the people gathered in St. Peter's Square, he had this to say:

> Among the voices in the world which we must listen to in order to obey the Council — and that means the spirit of the Gospel — there is, these days, that of the FAO, the international organization which studies food problems, which has its headquarters in Rome and is now holding its General Conference.
>
> What does this voice say to us? It tells us that half the population of the world is short of bread.
>
> This fact should produce among good persons many new proposals which the Council is awakening: to know these problems

*I expressed myself in a magazine article at that time in these words:

"A man is not a digit. He is a sentient being, gifted with immeasurable possibilities, whose fulfilment requires that he grows as an individual in a community of individuals. He must ask himself if the quality of life in a world with an infinitely expanding population will be in any way worth having. I believe that having done so, he will not elect for a future ruled simply by crude consideration of man's survival, but for one in which an optimum population goal takes heed of all the nobler attributes distinguishing man from beast — creative intellect, art, spirituality, the quest of self-knowledge, mutual consideration, compassion. Few of these virtues will be able to survive in a world sacrificing human quality for uncontrolled numbers."

> and to emerge from the state of indifference concerning them, to help those who are working to remedy them.
>
> The Church too is trying to do its part, by further developing its work of relief and charity. Moreoever, it does so by calling on Divine Providence, before anything else, so that it may help those who are suffering and those who are seeking means of promoting work and new plans to relieve hunger in the world, and so that it may inspire those who have the means to new acts of charity toward their needy brethren.
>
> With Our Sunday prayer We invite you all to this first recourse to Providence.

As time passed, Pope Paul's attitude towards population control became more traditional. In 1967, Pope Paul in his Encyclical *On the Development of Peoples* appeared to me to take a broader view on the question (Para. 37):

> It is true that too frequently an accelerated demographic increase adds its own difficulties to the problem of development: the size of the population increases more rapidly than available resources, and things are found to have reached apparently an *impasse*. From that moment the temptation is great to check the demographic increase by means of radical measures. It is certain that public authorities can intervene within the limit of their competence, by favouring the availability of appropriate information and by adopting suitable measures, provided that these be in conformity with the moral law and that they respect the rightful freedom of married couples. Where the inalienable right to marriage and procreation is lacking, human dignity has ceased to exist. Finally, it is for the parents to decide, with full knowledge of the matter, on the number of their children, taking into account their responsibilities towards God, themselves, the children they have already brought into the world and the community to which they belong. In all this they must follow the demands of their own conscience enlightened by God's law authentically interpreted, and sustained by confidence in Him.

I interpreted this directive to mean that this gave sufficient latitude to public authorities to promote birth control provided that they were of the view that it was in accordance with their interpretation of "moral law". The interpretation I gave to a Press Conference which was called immediately was hotly contested by the Representatives of the Vatican. I issued a press release which was carried by news agencies on what followed:

> Vatican City (UPI) — Pope Paul VI Monday discussed his encyclical *On the Development of Peoples* with the head of the UN Food and Agriculture Organization (FAO) but they apparently kept clear of their difference of opinion on its birth control provisions.

Informed sources said Pope Paul told FAO Director-General Dr. B. R. Sen of India he hoped the encyclical published last week would help promote speedy expansion of worldwide programmes to help the needy nations.

He also promised the FAO chief he could count on the Catholic Church's full support for FAO assistance programmes.

Vatican and FAO sources agreed that the difference of opinion between Pope Paul and Sen on birth control was not raised in the 20-minute private audience.

In his encyclical Pope Paul approved government promotion of birth control programmes providing these respect "moral law".

Sen, in a news conference Thursday in which he praised the encyclical highly, interpreted the Pope as meaning each country should be guided by its own moral law.

But a Vatican spokesman said the following day that Sen had erred. "Moral law is equal for all," Msgr Fausto Vallainc stressed. "It does not vary because it is natural law . . . this is one of our basic tenets."

In the Vatican's view, any government birth control programme should respect the Catholic ban on the oral contraceptive pill and other artificial means of birth control.

In Sen's view, government programmes could use such means if the moral codes of their own countries sanctioned them.

Sen was accompanied in Monday's audience by Msgr Luigi Ligutti of Des Moines, Iowa, and Dr. Emilio Bonamelli, the Vatican's permanent observers at FAO.

During the Ecumenical Council, Pope Paul appointed a Committee of Bishops to go into the question of population control. I was asked by their Secretariat to supply a lot of material relating to population growth and available food supply. I consistently made the point that it was not so much a question of potential resources which could be exploited to feed the growing population but of resources which were in actual fact being exploited at any point in time. I understood that the report of the Committee was in favour of population control in certain circumstances. Before I left Rome at the beginning of 1968 I happened to meet a close aide of Pope Paul at a dinner party given by Ambassador Count Schwartzenburg (Austria). I pleaded with him to convey to His Holiness my earnest wish that the Encyclical should be left alone for the world to interpret it. But another Encyclical appeared before long setting the matter at rest on clear traditional lines.

Both the Popes honoured me with their moral support in my conduct of the Freedom from Hunger Campaign and they both invested me with honours unusual for one who is not of the Catholic denomination. To me Pope John was one with divinely endowed qualities — a man of compassion and understanding. Pope Paul seemed to be torn between the ideal and the real. Despite our differences, I never felt deprived of Pope

Paul's goodwill. He was always approachable. In 1965, before he left for New York to address the United Nations General Assembly, I wrote to Cardinal Cicogniani, Secretary of State, to request His Holiness to mention in his address the problem of hunger in the world and the need to harness the world's resources to free mankind from the degradation of hunger and poverty. In reply, Cardinal Cicogniani said: "The Holy Father does not ignore the indispensable contribution that FAO brings to man to establish a necessary balance among the peoples, or the indefatigable activity of its Director-General. That is why he invokes with all his heart the celestial benediction on that agency and on your Excellency personally. I am happy to transmit this message."

Chapter 16

Sequel to the Campaign

Jumping on the FFHC bandwagon

SOON AFTER the World Food Congress, a Resolution (No. 1943 XVIII) was submitted by the UK to the General Assembly and adopted by its Second Committee, proposing a World Campaign against Hunger, Disease and Ignorance for the second half of the Development Decade. At the FAO Conference of 1963, the UK Minister of Agriculture, Christopher Soames, explained that one of the objects of the resolution was to enable the Secretary-General to consider, in consultation with the heads of the Specialised Agencies, how the resources of the United Nations' family might best be mobilised to stimulate and channel the efforts and goodwill of private individuals and organizations in such activity. Mr. Soames appealed to all non-governmental organizations now working for the FFHC to take up this wider campaign.

Only a few days before the FAO Conference, the National FFHC Committees had gathered in Rome and had called for the placing of all the National Committees on a continuing (or even permanent) basis. They expressed their opposition to the proposed take-over by the UN. My own view, which I expressed strongly at the Conference, was that whether the Campaign was carried on in its present form or with a widened mandate which would give greater emphasis to health and education, the essential factor for success was the involvement of people, supported by their governments, and guided and coordinated by such UN Organizations as might be concerned. The proposal to include the problems of illiteracy and health was a tribute to the work that National Campaign Committees and NGOs had done under the Freedom from Hunger Campaign. I said, however, that if the suggestion was that the very success of the FFHC should be the reason for dismantling it, so that its place could be taken by a similar but bigger campaign with another name, then I would have serious doubts about supporting it. The FAO Conference did not come to any conclusion on the General Assembly resolution, but at the same time resolved that the FFHC activities under the sponsorship of FAO "should be continued beyond 1965".

The next forum for the debate was ECOSOC at Geneva, since it had been asked by the General Assembly to explore the feasibility of organizing and executing a wider campaign. Before going to ECOSOC I had an opportunity of meeting U Thant in Paris at a meeting of the Heads of the UN Agencies (Administration Committee on Coordination/ACC) to explain my misgivings. I found him quite understanding. Maheu (UNESCO) was of course an enthusiastic ally in favour of the wider campaign, as education (or rather UNESCO) would then have equal prominence with FAO. In the report which he had been asked to submit U Thant was very cautious. He said that while there was some support for a wider campaign he did not think that the support was sufficiently widespread or specific. It was also his view that considerable weight should be attached to the reservations made by a number of governmental and non-governmental organizations on the disadvantages of proliferation of world campaigns, and he also noted the difficulties of satisfactorily solving the relationship between the proposed campaign and the FFHC. The dangers of launching a major new initiative without any reasonable assurance that it would be successful, he said, required no emphasis. The debate began. It was really a debate between me, on one hand, resolved not to allow the FFHC to be dismantled, and the UK delegate who was the main supporter of the General Assembly resolution. In my opening statement, I reminded ECOSOC of my initiative in 1958 when I introduced the idea of a Freedom from Hunger Campaign and that such a campaign needed the support of the entire UN family even though FAO might take the lead. This was the course that the campaign had taken. At the World Food Congress, most members of the UN family concerned had taken part. The Secretary-General in his speech had identified the objectives of FFHC with the Charter of the UN itself and of the UN Development Decade. The Director-General (DG) of UNESCO had said the campaign was first and foremost an educational campaign. The DG of WHO said there was an identity of purpose between the two organizations. The DG of the International Labour Organization (ILO) considered the campaign worthy of its fullest support in obtaining social justice. Thus from the very beginning, the campaign had not confined itself solely to food, but had encompassed the related aspects of education, health and other conditions essential for general economic development. I then outlined the way the campaign had proceeded and referred to the wide support it had received from world leaders. One of the most significant initiatives of this campaign, I pointed out, had been to enlist the effective participation of the people.

Against this background, I asked ECOSOC to judge how best to give effect to the General Assembly resolution. I expressed my own views in these terms: the strength of FFHC lay largely in its sharp focus on the concrete issues of hunger. Beside that, the campaign now proposed seemed too wide and indefinite in scope. What is meant by Ignorance? I asked. Lack of the three Rs or lack of general education? We know man's ignorance is and will always remain limitless, and is ignorance not a spur to knowledge? Again, does Disease include all diseases? If so, how can the general public be expected to campaign against them? I pointed out how the National Committees had approached their task. The UK National Committee had devoted nearly half of its funds to educational projects related to hunger and malnutrition.

There appeared to be two ways of implementing the General Assembly resolution — by enlarging the scope of the FFHC or by launching simultaneously a Campaign on Health and a Campaign on Education. So far as the first alternative was concerned, this was not only feasible, but also desirable and even essential. The title of a wider campaign, if decided upon, should still remain FFHC rather than be changed to a purely descriptive one. Also, it would be valuable to maintain our well-established channels of communication with the NGOs who were the main instrument for mobilising people's participation.

If, on the other hand, the GA's intention was interpreted as requiring equal emphasis on the whole range of health and education activities, the most logical course would be to have three independent campaigns operating side by side with each Specialised Agency taking its own responsibility and the coordination being provided through ACC.

In reply, the UK representative did not face the substance of my arguments but set about making a personal attack. In one outburst, he shouted, "When the King is weak, the Barons become strong!" suggesting, I suppose, that I was manipulating the Secretary-General. After his speech, the veteran UN Assistant Secretary-General, David Owen, came to me and said that he had never seen such a personal attack in a UN debate.

ECOSOC came to the conclusion that "the existing circumstances were not favourable to the launching of such a campaign".

Chapter 17

FAO and World Issues

On Disarmament

DURING THE Marshall Plan operations, the USA spent as much as 5 to 6 per cent of its GNP in aid to the war-devastated countries of Europe. It was now clear that to help the developing countries break out from their economic stagnation within a reasonable time, aid in different forms would have to be on a Marshall-Plan scale far beyond the trickle forthcoming. I often thought of what could come out of a worldwide scheme of disarmament if it ever became a reality. In July 1963, the news came through of the Test Ban Treaty initialled in Moscow by the USA, USSR and UK. Immediately, I gave the following message to the Ambassadors of the three countries:

> The Treaty initialled in Moscow on nuclear tests has filled the world with expectation. The peoples of the world are hopeful that this Treaty will open a new era of international understanding, a new era of lasting peace and security. They are thankful to those who have made this historic agreement possible.
>
> But the attainment of peace is a continuing and positive process. It requires that all Nations shall cooperate in creating conditions of true security. This security cannot be attained as long as more than half of mankind is not freed from the curses of hunger, malnutrition and poverty.
>
> As early as 1942 President Roosevelt proclaimed Freedom from Want as one of the four objectives to be achieved through international cooperation. He realized in full the major importance of

this factor of world peace and the need for Nations to dedicate a major share of their resources, both human and economic, to this joint, constructive endeavour. Since then, significant advances have been made in the fields of science and technology and the knowledge needed to eradicate hunger and want is available to us.

More recently, FAO's Third World Food Survey has shown the true dimensions of the problem of hunger and malnutrition which exist in the world. The late Pope John XXIII in his Encyclicals, particularly in *Pacem in Terris* as well as in his messages supporting the Freedom from Hunger Campaign, set out the immediate duty of society in this respect. The Manifesto issued in March by a Special Assembly of Nobel Prize Winners and other eminent world citizens emphasised what even a partial diversion of military expenditure could achieve. In June this year, the World Food Congress in Washington adopted a Declaration concluding with the hope "that the current efforts for bringing about universal disarmament will succeed and that the vast sums now being spent on instruments of destruction will become increasingly available for the elimination of hunger and malnutrition and the promotion of human well-being.

These declarations have aroused hopes everywhere. The creation of conditions that will make life worth living throughout the world is a challenge for all Nations, rich and poor. As the Chief Executive of the Food and Agriculture Organization of the United Nations I regard it as my duty to appeal to the Heads of the Governments represented in Moscow to include as a specific provision in their agreements the establishment of a World Disarmament Fund:

(a) to come into operation immediately,
(b) to be financed by contributions from all members of the United Nations and the Specialised Agencies, linked to the savings emerging from reductions of military expenditures, starting with the relatively modest amounts resulting from the suspension of nuclear tests,
(c) to be devoted to the worldwide achievement of Freedom from Hunger and Want in our time.

The replies were, as expected, very cautious; only the USSR was a little more forthcoming. I reproduce the USSR's reply in French which accompanied the original telegram.

Le message de Monsieur Sen au Président du Conseil des Ministres de l'URSS Nikita S. Chrouchtchev envoyé par l'intermédiaire de l'Ambassadeur Soviétique à Rome est transmis à sa destination.

Le Gouvernement Soviétique attribuait toujours et attribue à présent une grande importance au problème de l'aide aux pays et peuples qui ont tout récemment pris la voie du développement indépendant et qui ont une économie sous-développée. L'Union Soviétique dans une mesure grandissante prête aux nombreux pays de l'Amérique Latine, de l'Asie et de l'Afrique son aide économique et son assistance technique à la création de la base d'une économie nationale indépendante et à la liquidation de l'héritage pénible du colonialisme.

L'Union Soviétique est de l'avis que les ressources humaines et économiques qui seront libérées grâce au désarmement doivent être utilisées pour le développement des secteurs pacifiques de l'économie et pour l'augmentation du niveau de vie des peuples. Quant aux méthodes et aux formes de l'utilisation de ces ressources l'opinion est qu'il serait plus raisonnable examiner ce problème après la conclusion de l'accord sur le désarmement. A présent le but principal consiste à atteindre le plus vite possible une entente entre les états sur la réalisation des mesures du désarmement et de la libération des moyens consacrés jusqu'à présent aux dépenses militaires. Le Gouvernement Soviétique estime que les Nations Unies et leurs organisations spécialisées pourraient prêter un concours considérable à l'obtention de ce but.

It was quite clear to me from the US Embassy's reply that my message never reached the White House, and might not even have been put before the Secretary of State, following the usual practice of the lower echelons of the US Administration (as I learnt from my Washington days) of disposing of important matters on their own until a formal protest was made. The US *Chargé d'Affaires* wrote in reply:

> The United States appreciates the humanitarian spirit which motivated your proposal ... agrees thoroughly that as and when disarmament really gets underway, the savings in part can be used to supplement the resources which are otherwise available for development purposes. In the meantime, however, the US considers that it would be unwise to give to those who oppose development an excuse for doing little or nothing now. There is danger that public statements by world figures linking disarmament with development might have a reverse effect.

I was asked to treat this reply as secret. I am inclined to think that one reason for this request for secrecy might have been that if my message had attracted the attention of President Kennedy, the reply might have been very different. It is a pity indeed that in recent years this possible linkage between Disarmament and Development has been lost, overtaken by other "linkages" of more immediate political concern.

Apart from this attempt to interest the Nuclear Powers in the problem of World Hunger, the pressure on world public opinion was maintained all through the Campaign in other ways. I have already referred to the Manifesto issued by the Nobel Laureates. I have also quoted from the Declaration of the World Food Congress held in Washington. In 1965 I invited Youth Leaders from all the Continents. In their Manifesto they arraigned the present world trends: "The world is ruled mainly by people out of touch with the young world. They know that men starve and die in millions. But they think it more important to make guns, bombs, warships, rockets and send us to fight one another than to provide seed and water, schools and hospitals, so that we might feed and serve one another."*

Yet the arms race continues unchecked and, in fact, it is on the increase. The global competition now consumes over US$400 billion a year in public funds. The Swedish Institute SIPRI estimates that over a billion dollars a day, or over a million dollars a minute, are being spent in arms of various kinds, a disproportionate part of it in the Third World. No doubt, this is a reflection of the tensions in the world today. Rising conflicts of ideology, re-emergence of competing geo-politics (which is only another name for the old familiar concept of "spheres of influence"), attempts to readjust political boundaries by war in the aftermath of the withdrawal of Colonial Powers and the purchase of more and more arms of high technology by the fabulously wealthy Arabs — all are factors which put our civilisation on the brink of annihilation. But where does the ultimate security of a nation lie? In piling up arms at any cost, or in trying to enrich the lives of the people through better health and educational facilities, and providing greater opportunities for productive employment? What would a half, a third, or even a quarter of the yearly arms expenditure do if used for development? One remembers the wise words of Spinoza, the philosopher. "Peace", he said, "is not just absence of war; it is a virtue, a state of mind, a disposition for benevolence, confidence, justice..."

On Hunger and Human Rights

I FIRST MET Eleanor Roosevelt in the Third Committee of the UN General Assembly debating the draft Declaration on Universal Human Rights. I was one of the Indian delegates. She was one of the architects of the Universal Declaration. I was a silent, respectful listener. Several years later, when I was Ambassador to the United States, my wife and I had the privilege to be invited by that great lady to her Hyde Park home. She took us out into the garden to show us the tomb of Franklin D. Roosevelt. She pointed out the space reserved for her beside him, when that time would come.

In December 1948, the General Assembly proclaimed the Universal Declaration of Human Rights. In the early sixties, the question of

*See Appendices IV, V and VI (pp. 314-321), respectively.

drawing up covenants in order to make the principles of the Universal Declaration more legally binding was raised in the General Assembly. The year 1963 was the fifteenth anniversary of the Universal Declaration. This was also the year when the world's attention had focussed on our Congress inaugurated by President Kennedy in Washington. I thought that the timing and the climate were just right for me to appear before the General Assembly to plead for the inclusion in the Covenant of man's right to freedom from hunger. The draft Covenant when drawn up, in Article II.2, had the following provisions:

(1) The States Parties to the present Covenant recognize the right of everyone to an adequate standard of living for himself and his family, including adequate food, clothing and housing, and to the continuous improvement of living conditions. The States Parties will take appropriate steps to ensure the realization of this right, recognizing to this effect the essential importance of international cooperation based on free consent.

(2) The States Parties to the present Covenant, recognizing the fundamental right of everyone to be free from hunger, shall take, individually and through international cooperation, the measures, including specific programmes, which are needed:
   (a) to improve methods of production, conservation and distribution of food by making full use of technical and scientific knowledge, by disseminating knowledge of the principles of nutrition and by developing or reforming agrarian systems in such a way as to achieve the most efficient development and utilisation of natural resources;
   (b) taking into account the problems of both food-importing and food-exporting countries, to ensure an equitable distribution of world food supplies in relation to need.

I brought the draft Covenant to the notice of the FAO Council in 1965. The Council agreed that "The Preamble of FAO Constitution should not on such a problem be any less specific than the proposed Covenant." The recommendation of the Council was accepted by the Conference (1965) and the final words "and ensuring humanity's freedom from hunger" were added to the Preamble. A formal ceremony was held during the Conference to unveil the marble plaque at the entrance to the FAO building, setting out the Preamble to the Charter as amended.

It took ten more years to obtain the thirty-five ratifications necessary to bring the Covenant into force. On October 5 1977, President Carter gave his signature on behalf of the United States.

For me, the amendment to FAO's Constitution and the inclusion of man's right to freedom from hunger in the UN Covenant of Human Rights remain two lasting mementoes of the efforts we had made to bring before world consciousness the plight affecting millions of our fellow human beings.

Chapter 18

Planning for Development

The Mediterranean Project

AN INTEGRATED APPROACH to regional and global planning for long-term economic and social development has over the years been one of the greatest contributions FAO has made. The United Nations is preparing to embark with confidence on a study of *The Future of the World Economy* with the year 2000 in mind. Those responsible for this decision must be encouraged by FAO's pioneering studies, in particular *The Indicative World Plan for Agricultural Development*, taken up and completed on the recommendation of the 1963 World Food Congress.

Our first effort in planning for development was what we called the Mediterranean Project. The origin of this project lay in the series of studies undertaken by the Executive Secretary of the European Economic Commission, Professor Myrdal, which concentrated on agriculture, forestry and general economic development of the Mediterranean region. When he was unable to proceed further for lack of resources he asked FAO to complete his project. He sent me the results of his work up to that point and sent one of his young experts, Henry Ergas, to meet me. Ergas later proved to be invaluable. I decided to make a preliminary survey of the situation in the region on which I would then build a strategy of development. Among the main issues highlighted by this preliminary survey were: the need to stimulate a self-sustaining growth; the need to increase the flexibility of the economic and social system through institutional changes; the way foreign aid could be made more effective; and how the impact of UN and bilateral technical assistance might be increased. It was clear that a broad and bold attack on the entire problem was needed. This was beyond what FAO alone could do. I asked for the assistance of UN economic experts in studies to be called in. I also decided, because of FAO's limited financial resources, that of the ten proposed country studies three would be prepared by the countries themselves but in continuous consultation with us, and the other seven, along with the overall report, would be FAO's responsibility. To make sure that this study would not simply deal in theoretical possibilities I stipulated that the studies should be regarded as exercises in national planning. I wanted the final document to influence the policies and programmes of national governments. With this in mind I asked the ten countries to establish National Committees consisting of representatives of the ministries and other agencies concerned.

The Committees were set up in all ten countries, and the FAO team met with them and reached agreement on the general lines of the study. The overall report, based on these country studies, attempted to give a broad regional perspective to assist national planners. At the same time I made a special point to help the governments to implement the programmes. The broad conclusions of the overall study were:

(a) The rapid population growth could within fifteen years add about 70 million more people to the present population of 175 million.

(b) The neglect of a long-term comprehensive approach in the past had led to a situation of excessive deforestation and soil deterioration, denuded hillsides, advancing steppes and deserts. The static economy had prevented countries from responding readily to the changing structure of world trade. Foreign financial and technical aid, also granted on a short-term basis, had not been adequate to reverse trends.

(c) The dominant role of agriculture and therefore the need of concentrated financial and technical aid from abroad.

(d) The striking similarities in the problems of all these countries in production patterns, in lack of water, in excessive pressure on land of limited productivity, and the belief that these countries could gain from broader economic cooperation and in adoption of commercial policies.

It was also clear that a programme of this size would call for major institutional changes and for active development policies. The application of such a programme would raise difficult questions of choice, priorities, and balance, since the whole development process would consist of closely linked action in diverse fields and departure from some long-established ways. To answer the needs of the growing population, one would have to promote a policy that would have less and less people dependent on unproductive land and finding employment for them in industries and in public works. The latter might prove to be very important if works such as the construction of dams, reforestation, improved forest management and road building were taken up. Public works of this kind could provide the basic infrastructure, train people in future jobs and secure balanced growth by providing markets for increased industrial and rural supplies.

It was clear that a piecemeal approach to stimulate certain sectors without due consideration of overall implications could not be justified and that only after a broad analysis had been carried out would governments be in a position to select not only those sectors with a propulsive role, but also what were called the spearhead areas where initial efforts would be undertaken to spread outwards in the course of time. But while attention should be given to industrialisation the basic effort had still to be focused on agriculture. The most serious consequence of neglect of agriculture would be that it would leave the bulk of the population outside the process of development.

The implementation of programmes derived from this study would impose serious strains on the peoples concerned. The institutional changes would have to be accepted willingly by the peasantry and not simply be imposed on an authoritarian basis. A considerable overhaul of agricultural education, and particularly of extension work, would be necessary. Other reforms would include administrative machinery, greater equality of income through progressive taxation policies, land reform and cooperative farm credit, together with more dynamic fiscal and monetary policies.

For these countries, financial and technical aid would not be enough. A process of self-sustaining growth could be generated if they were to find profitable and expanding markets for their products. To achieve this they would have to adjust their commercial and commodity policies.

From 1957 to 1959, we were engaged in the general and country studies. At that time we had great difficulty convincing people of what we called the Mediterranean approach of integrated development. From 1959 to 1961 we attempted to define concrete follow-up action. In 1961, we held a conference in Badajos, Spain, and all the countries concerned took part. We discussed the establishment of spearhead development zones. The progress of the spearhead zones was discussed further at Nîmes, France, at a conference in May 1964. The Nîmes Conference also made a general survey. Among the important points brought up were:

(a) Marketing the same products internally and externally was an important problem and an appraisal of the market potential in relation to the production plans of the region as a whole was necessary.

(b) The importance of a continuous dialogue between those responsible for planning the development zones and the policy-makers at the centre was stressed. Indeed, that was why the field teams were conceived from the very beginning as joint teams made up of FAO experts and local counterparts. The Governments were asked to assume full responsibility both for planning and implementation.

(c) Much stress was laid on human investment — improvement of living conditions and new approaches in for example the training of skilled manpower.

During the five years 1959-64 and beyond, much careful work was done in some countries. Many lessons were learnt both by us and the countries involved. Some of the important "Special Development Zone" projects carried out in cooperation with the UN Special Fund (UNDP) included the Rif and the Sebou projects in Morocco, the Western Peloponnisos project in Greece, the Antalya project in Turkey, the Ghab project in Syria, and the project for the development of the Lebanese highlands. These were in addition to many national projects carried out in France, Italy and Spain. The project also provided occasions for the meeting of experts of international standing among whom may be mentioned Professors Gunnar Myrdal of Sweden, Balogh of Oxford University and Baade of Keele University, assisted by high-level national administrators like Philippe Lamour, President of Compagnie Nationale de la Region du Bas Rhône et Languedoc, France, Ahmed Ben Salah, Minister of Planning and Finance of Tunisia, and General Subeih, Chairman of UAR's Desert Development Authority. For me it was a most educative exercise. I came into personal contact with some of the basic problems of development, and realized the value of some of the new and rather unconventional approaches brought out during the study. Unfortunately, the whole political scene in the Middle East became chaotic, volatile and uncertain with the Arab-Israeli war. I hope that one day, when peace has returned, the eleven volumes of this pioneering Mediterranean Project will be taken down from the shelves where they are gathering dust and that work will once again begin with fresh vigour and hope.

Two persons to whom we owe a special debt of gratitude for this study are Thomas (now Lord) Balogh, who had a sharp mind and an equally sharp tongue (which made him unpopular with my colleagues), and Egon Glesinger of the Forestry Division who provided the ballast, and sometimes the direction, for a ship that was sailing through uncharted seas.

Indicative World Plan for Agricultural Development (IWP)

THE INDICATIVE WORLD PLAN for Agricultural Development (IWP) was FAO's most important pioneering exercise in global planning. It has since led to various perspective studies of a comprehensive nature in the UN family.

The final report was based on an analysis of the major issues that would confront the role of agriculture in the decades to come, and it set out the directions for national and international action. It embodied the results of four regional studies — the Middle East, South and Central America, North-west Africa and Africa south of the Sahara, and Asia and the Far East.

In undertaking this pioneering exercise, we started with some basic assumptions. Of these, one was the medium variant of the UN Population increases; another was the "feasible and sustainable" growth rate of the gross national product (GNP) for individual countries.

As a substantial base for projecting demand for food and other agricultural products as well as for supplies and investments for the years 1975 and 1985 — the two years set as target dates for achieving the objectives of the Plan — we used 1961-63, the latest period for which reasonably complete statistics were available. The Plan was not intended to be a detailed blueprint for immediate implementation; the detailed planning was to come from individual countries. The Plan was intended to assist the countries to set their particular problems within the world framework and determine their own course of action.

When the idea of such a world plan was first brought to my notice by Henry Ergas from one of the Commissions of the World Food Congress I believed that the idea required no more than an adjustment of *SOFA* (FAO's annual publication on the State of Food and Agriculture), and putting together and updating material from the different sections of FAO. I also thought it would be an extension of our Mediterranean Study. I agreed to the plan and asked Ergas to put the suggestion in a more concrete and positive form. The proposal was presented as a world plan, in quantitative terms, on the basis of nutritional needs and indicated the type and magnitude of internal effort and adjustments called for, as well as the type of external assistance needed in relation to local resources. It was to be internationally coordinated. When I returned to Rome and gave more thought to the proposal, I realized that we had taken on a tremendous commitment which would be difficult to carry out without a total reorientation of the work of FAO, both at Headquarters and in the field, and would require a sizeable increase in our budget. However, the proposal was challenging and I braced myself to face it.

I first referred to my preparations in a statement to the FAO Council in 1964. I said that time would be needed to develop such a Plan. At first we would only be able to give a rough approximation of the magnitude involved, in terms of production, consumption and trade targets, and the measures, including investment and trained manpower, necessary to achieve them. Such a global framework would gradually be filled in as more and better data became available. I was going to make specific provisions in the budget (1966/67) for a hard-core group (under one of my staff members, Walter Pawley, a trained economist who had a wide vision of the problems) to work on this Indicative Plan. Further supporting activities would require selective strengthening of the Regular Programme activities at a number of points. At the FAO Conference following the World Food Congress, I had said that the survey should take the form of a World Plan for agricultural production, trade and development, which would highlight the national and international actions necessary.

By the next FAO Conference in 1965, ideas had already crystallised and I expressed my conviction that the Plan might for the first time provide a focus for the wide spectrum of FAO's responsibilities, as well as an international frame of reference for our member countries in their national planning. It could also furnish a framework within which donors and recipients could see more clearly the priorities in respect to external aid. The Plan's immediate horizon would extend to 1975 with a longer-term perspective to 1985. For these dates, the Plan would seek to arrive at consistent world and regional targets for consumption, production and trade of agricultural commodities, based on the assumption that efforts would be made to reach the growth objectives of the world economy as projected by the UN. The analytical foundations for such work had already been laid through its long-term projections studies. Naturally FAO would have to further develop the methodology and techniques as the work proceeded. Nonetheless, the dual approach should prove fruitful: on the one hand, looking at the world outlook for the main agricultural products, and on the other, building national data into consistent aggregates for natural or economic regions. Indeed the examination of potential intra-regional complementarities and trade in the agricultural sector could well prove one of the most useful aspects of the work.

I stressed that the Indicative Plan would not be a set of projections and targets only. To achieve its purpose, the Plan had to provide a reasoned guide to action by governments, made in the context of the analyses presented by the Plan, which alone could result in the attainments of the global targets. Thus, one of the objectives of the work would be to bring into sharper focus some of the key policy issues faced by the developing countries in their planning. I cited a few selected problems to illustrate

capital goods must be imported. With acceleration of economic growth, investment requirements rise and so also do requirements for foreign exchange. There is therefore a need for a major increase in earnings from agricultural exports. Fluctuations of price have a specially serious effect, since many of the developing countries are dependent on exports of a very limited range of commodities. Fluctuations in foreign exchange seriously impede development plans of these countries. Thus radical action is needed to make international trade a factor of growth and prosperity rather than of risk and uncertainty.

Trade and aid must be coordinated to complement each other in supporting development efforts. It was tragic that many of the developed countries put up artificial barriers, protectionist as well as fiscal, against imports from the developing countries. On the other hand their exports were often subsidised or given other forms of assistance. In the situation that we faced the concept of regulation of trade by free market forces was no longer tenable. Those conditions called for a fundamental change in policies and marketing structure. The international commodity agreements must be wider than mere stabilisation of prices over a long-term period, though that was essential. They must include policies influencing production, domestic price levels and related commercial policies. Proposals had been made for systems of levies on imports of agricultural commodities which would go towards building up development funds and a portion of this money could be allocated to those countries exporting a given commodity and the rest would be kept as general capital aid. But such payments would not form part of the prices received by producers, and thus would not act as incentives to increased production. Within the definition of Aid, we could include:

(i) long-term loans repayable in foreign currency,
(ii) grants and "soft" loans,
(iii) grants or sales of surplus products for local currencies,
(iv) technical and pre-investment assistance generally. (This excluded private foreign investment which formed by far the greater part of foreign capital inflow.)

UNCTAD-I therefore had to give attention to measures to increase foreign investment. The developed countries might consider the establishment of an Insurance Fund to protect and encourage outflow of their private capital but without the intervention of foreign policy considerations; while the developing countries could see what measures might be taken to create a favourable climate for foreign investment. The scale of aid should be increased. It was now far below the one per cent recommended by the General Assembly. A positive incentive for increased national effort would be forthcoming if it was believed that all requests which met functional criteria of productivity would be granted. Knowledge that capital would be available over a decade or more, up to the limits of the capacity to absorb it, would act as an incentive to greater effort. The kind of national development plans which specially called for international support needed to be stressed. Such a national development plan must be conceived as a charter of policy which outlined not only what the people might expect in time, but also the duties and the sacrifices which they might undertake. It should not only give a clear picture of the economy and the possible targets, but also describe the institutions and policies designed to ensure proper implementation. The more clearly the whole pattern was coordinated the wider would be the limits to which demand might be permitted to expand and, consequently, the larger the amount of aid that could be utilised.

I referred to the importance of the UN family providing incentives to national effort to draw up national plans, to undertake resource surveys, pre-investment studies and major educational projects. I stressed the importance of technical assistance at key points, of loans on easy terms (the World Bank and International Development Authority) and of agricultural surpluses for economic development. I referred to the growing reluctance of some of the developed countries to increase their contributions to the budgets of the UN Specialised Agencies, when the developing countries were hoping for a more rapid expansion of the programmes. Since bilateral assistance was also decreasing on the pretext that the load should be more equitably shared, it seemed to me that there was an urgent need to channel more resources through the assessed budgets of the UN family. The lack of progress so far could not be ascribed to any lack of institutional arrangements, methods or machinery. It was political will rather than new machinery that was required to meet the situation. There was often a tendency to look upon trade as something apart. Trade problems in agricultural products were intimately related to production, consumption and general economic development.

For these reasons, specific responsibilities had been placed on FAO. The studies, consultations and action in this field by FAO formed an essential part of international efforts towards a balanced expansion of the world economy. FAO was also charged with promoting national and international action to improve processing, marketing and distribution of food and agricultural products, including fisheries and forestry. Formulation of recommendations for international policies on commodity arrangements was also a specific FAO responsibility. Another important FAO task was to undertake an ongoing review in the hope of promoting action related to the sound development of prices and trade. FAO's Committee on Commodity Problems (CCP) had been particularly successful in promoting intergovernmental consultations on problems of individual commodities through Commodity Study Groups, and in developing agreed policies and procedures for the disposal and utilisation of surpluses for economic development. FAO services in this respect were utilised both within and outside the UN family. FAO stood ready to strengthen its own machinery to make the maximum contribution. The problems were indeed difficult and complex. Yet one fact stood out. Unless the high-income industrialised nations agreed to open their markets at remunerative prices to the products of the developing countries, the hopes that this conference had aroused would be betrayed.

The validity of any of these arguments has not been challenged in the last fifteen years — but has rather grown in strength in UNCTAD.

One recollection stands out from that conference — my meeting with Che Guevara who came at the head of the Cuban delegation. The meeting was arranged through a mutual friend. I was keen to meet the great revolutionary. As he came in, I was struck by the gentle and rather shy smile on his handsome face. We talked of inconsequential things, but I was fascinated by the expressions on his face. A short time later he was to be hunted down in Bolivia and shot. What a wasteful end to a precious life — so young, so full of hope!

Trade between Developing Countries

THE PROBLEM of promoting trade *between* developing countries first came up as an issue before the international organizations out of the studies made under the Indicative World Plan. In these studies detailed examination was made of the possibilities of trade between, for instance, Middle Eastern and East African countries. Suggestions were made on the basis of complementary production and trade between them. I put the issue before the Council in the following terms in October 1966:

> The structure of commercial trade today still reflects the realities of fifty years ago. Well-structured and efficient trading channels exist almost exclusively between high income and low income countries. In one direction they have been coping with large exports of primary products, and in the other with the imports of equipment and consumer goods that are not produced in developing countries. So long as this trading structure remains unchanged, the import requirements of developing countries would have to be met largely from high income countries, either by trade or aid.
>
> At the same time the prospects for the traditional export commodities of developing countries continue to be poor. It is essential that the net agricultural import needs of rich countries would at the very best increase by 2.5 per cent per annum in value terms, but this will require some major changes of policy as well as rapid economic expansion on their part. Faster growth cannot be achieved in developing countries without a dynamic export sector. There is an urgent need to find new markets and commodities for export.

A huge potential market exists among developing countries themselves for food and other agricultural products. A number of developing countries are already food exporters and several have the production potential to produce and export surplus. If trade between developing countries could be built up on a large scale, it would provide the vital stimulus necessary for faster growth in producing countries. But, with the present trading structure, there is little hope of substantial trade development on these lines. Scarce resources of foreign exchange further incline the needy countries to the present trade structures, looking for food and better credit terms.

Probably one way of stimulating large-scale trade flows between developing countries would be through setting-up a new clearing system, based on some kind of multilateral barter, or on a unit of exchange negotiable only between the developing countries. We have seen an arrangement of this type between the countries of western Europe immediately after the War. But the capital backing and international commercial knowledge of the rich would be needed to launch such an effort, though only concerted action of developing countries themselves could ensure its success. The implications of such a new Clearing House system for individual countries and commodities will have to be carefully studied.

Up to the end of World War II, the pattern of trade was a reflection of the needs and requirements of the Colonial powers. Shipping, banking and other interests followed the Flag. The problem of trade between developing countries became a question of giving economic content to the political independence attained by those countries. The Conference in Buenos Aires in 1978 called by the United Nations on Technical Cooperation between Developing Countries (TCDC) was a natural development following independence. But its title showed that its scope was limited to one facet only — technical cooperation. There was however a much wider field in which cooperation could be extended. Some of the developing countries had acquired some shipping to carry their trade. It seemed this could serve as the spearhead for all the other ancillary and supporting activities needed to give these countries real economic independence and help them to break out of the straitjacket inherited from their colonial past.

Chapter 20

The World Food Programme

A United Nations success story

THE WORLD Food Programme (WFP) is a success story for the United Nations family. Following parallel resolutions[1] of the United Nations General Assembly and the FAO Conference in 1961, the World Food Programme was launched formally on 1 January 1963 under UN/FAO auspices and located at FAO Headquarters in Rome. Intended in its initial stage to meet emergency food needs, it was also intended to assist countries in their economic and social development.

Since its creation as an experimental three-year programme, WFP has been further developed. After the first two years of its operation, the FAO Conference and the General Assembly put the programme "on a continuing basis for as long as food aid is found feasible and desirable".[2] Though in theory the WFP is working towards its own abolition in the hope that the need for multilateral food aid will, at some stage, no longer exist, its dissolution is nowhere in sight, given the state of economic and nutritional needs in the world.

Conceived as a programme for surplus utilisation, WFP uses food as an additional source of development capital through food-for-work and nutritional development projects, and it also helps to meet emergency food needs in times of disaster and crisis.

[1]General Assembly Resolution 1714 (XVI), adopted on 19 December 1961; FAO Conference 1/61, adopted on 24 November 1961. See Appendix III on Page 309.
[2]General Assembly Resolution 2095 (XX), adopted on 20 December 1965; FAO Conference Resolution (4/65), adopted on 6 December 1965.

In the sixteen years since its inception, WFP has committed US$4,000 million in food aid for economic and social development and nearly $500 million in emergency aid.

Today WFP covers a wide range of operations. More than two-thirds of its assistance goes to projects of agricultural and rural development. These include land improvement schemes with an emphasis on bringing wasteland into cultivation by irrigation, reclamation and anti-erosion works; dairy and livestock improvement; community development including building of houses, schools, roads, wells, water supplies, drainage, and large forestry programmes. It has also occasionally helped to mine coal, repair and maintain railroads, repair electric power lines, and even helped to raise productivity in factories. Growing emphasis has been placed on projects of nutritional improvement through school-feeding programmes and mother-and-child health centres.

I may cite some of the examples of the achievements under a unique approach which characterises the World Food Programme.

In the Egyptian plains which were once subject to flooding before the construction of the Aswan Dam, thousands of farmers are now raising two to three crops a year instead of a single crop. WFP's contribution has been the construction of a network of canals and tile drainage projects which has helped to add to their earnings.

In South Korea, the farmers trebled their earnings from crops by building flood protection dykes through a food-for-work assisted project. The biggest such project is in Bangladesh where more than two thousand miles of canals and embankments have been completed by men and women — up to two million of them at times — working just for food in the non-rainy season, people who were normally only employed on a seasonal basis.

In India, WFP has supported the biggest dairy development project in the world by providing butter, oil and milk powder to increase the milk supply by fifty per cent to four major Indian cities in a largely vegetarian country.

In Botswana and Lesotho in southern Africa, almost every school-going child receives a wholesome lunch in school through school-feeding projects. The children now show a better attendance record and a much improved school performance. In Colombia, one out of twenty people in the country is covered by a national nutrition improvement programme which is supported by WFP.

Development is a slow process, but food needs in the wake of disasters have to be treated on an emergency basis. The Director-General, through the World Food Programme, has been able to provide food worth more than $500 million to meet emergency needs in one hundred countries.* The aid has gone towards relieving conditions caused by severe droughts which occurred twice in the Sahelian zone in Africa, in Ethiopia and more recently in India, Afghanistan and Nepal.

*Figure quoted is as of end 1979.

Emergency food has made the difference between life and death in civil strife in Nicaragua, Lebanon, Nigeria and Kampuchea. Victims of earthquakes and hurricanes in Caribbean countries and Bangladesh, and victims of earthquakes in Iran and Yugoslavia who benefitted from WFP aid, are a testimony to the efficacy of this programme. I cite only a few of the hundreds of examples.

Taking action to meet emergency food needs was, from the beginning, a responsibility which fell to FAO under ECOSOC and FAO Conference resolutions. The Director-General of FAO was also required to keep under constant surveillance existing food shortages in individual countries, to assess the extent of international assistance needed, and to devise practical lines of action for "prompt, concerted and effective assistance". However, the Director-General had no specific resources at his disposal; he could only issue a general appeal to interested governments. When WFP was created, I sought to rectify this by including in the terms of reference of WFP a provision that specific food resources should be placed at my disposal for such purpose. This found great appeal and some leading delegations, notably the delegation from Canada, expressed the wish to have the entire programme resources devoted to emergency needs. It was at my insistent plea, however, that a balance was struck between the two purposes — emergency relief and economic and social development.

In some respects WFP is a unique enterprise. Being multilateral, it does not take foreign policy considerations into account when giving aid, a factor which unfortunately sometimes influences bilateral aid. It provides them in effect with "budgetary" support in terms of food which is equal to one-half of the wage bill in labour-intensive projects. It contributes to nutritional improvement as the rations are given to the worker and his family. All development expenditure has an inflationary effect on the economy. Insofar as the wages are paid in food, the aid helps to cushion this impact. For the donors, especially with large surpluses, food aid is easier to give. In any case it creates no balance-of-payments problems for them to the extent that their contributions are in food.

There are no fewer than a hundred donors to the World Food Programme. When it is realized that there are not a hundred rich countries in the world, this global cooperative effort falls into perspective. These contributions could be tea from India, rice from Pakistan, cocoa from Ghana, coffee from Colombia or small cash pledges from countries like Nepal and Cyprus.

The Programme is maintained by voluntary pledges from member countries of FAO and the United Nations. Two-thirds are in food commodities (more than half in wheat, maize, sorghum and other cereals) and one-third in cash and services such as shipping. The pledges are made to meet targets set by FAO and UN. Twice — in 1969-70 and in 1975-76 — pledges exceeded the target by more than 50 per cent. During other periods they have fallen short of the target. For 1981-82 the target is US$1,500 million.

History of the World Food Programme

WHEN THE Hot Springs Conference was held in 1943, at the height of the War, the world was facing acute shortages. But the Quebec Conference in 1945, while drawing up the Charter for FAO, was already anticipating accumulation of surpluses. It stated "It would be folly to disregard the possibility of surpluses, and FAO should study how to deal with such surpluses . . .". But when the question of commodity arrangements came up, FAO was asked not to get involved with commercial and commodity policies. The USA was particularly sensitive to any interference with their normal channels of trade. These two conflicting trends of thought have persisted ever since.

The first bold unconventional approach to the whole matter was made by FAO's Director-General, Boyd-Orr, in his proposals for a World Food Board. The Board was to:

(i) stabilise prices by holding buffer stocks;
(ii) deal with emergencies, such as famines, by holding a world food reserve; and
(iii) finance surplus disposal to food-deficient people.

Each of these called for international funds on a large scale, and surplus disposal did not guarantee non-interference with normal channels of trade. Boyd-Orr's proposals, somewhat ambitious and visionary for that time, were rejected and buried in a report from the Preparatory Commission (1946-47). According to the Commission, the objectives of his proposals could be achieved only in an expanding world economy which required the industrialisation and agricultural development of the underdeveloped countries. The report went on to state that price stabilisation could be brought about through intergovernmental commodity agreements; that exporting countries should arrange concessional scales of surplus commodities for nutrition schemes and finally, pending the establishment of an International Trade Organization, that an Interim Co-ordinating Committee for Commodity Arrangements should be established.

Another attempt was made to see if the accumulation of surpluses in hard-currency countries — the dollar area — could be disposed of by accumulating inconvertible currencies to the credit of the selling countries until general convertibility was restored. The 1949 FAO Conference rejected this proposal but established a Committee on Commodity Problems (CCP) to advise on difficulties arising out of surpluses. In 1952 this Committee examined the feasibility of establishing an emergency food reserve and suggested three possibilities:

(i) that a stock of food be held by an international agency;
(ii) that a fund be administered by an international agency for purchasing and distributing food; and

(iii) that emergency food reserves be held by national governments for international use.

None of these possibilities were found acceptable to the FAO Conference. When surpluses became an international problem again in 1954 and 1955 and the USA found it necessary to export on concessional terms, CCP set up a working party which examined the whole problem of surpluses and drew up a Statement of Principles to be adhered to by exporting countries to ensure that surplus disposals did not interfere with the normal patterns of production and trade. These Principles are still in operation which is to the credit of CCP. The CCP Working Party was later given the status of Standing Consultative Sub-Committee on Surplus Disposal with a mandate to oversee the operation of the Principles.

As for national food reserves, both the General Assembly and ECOSOC had asked me to make a special study of the question giving particular attention to the possibilities of using surpluses for building up reserves in underdeveloped countries. Our own Conference had also stressed the importance of adequate national reserve policies while at the same time taking into account the FAO Principles of Surplus Disposal. While preparatory work was undertaken both at FAO Headquarters and by the Washington Sub-Committee, I asked Dr. Gerda Blau, one of our Principal Officers (a highly trained economist and analyst) to make some field enquiries so that I could learn more about the problem. The report prepared by FAO following these enquiries brought out three main factors:

(i) The main remedy against both chronic want and excessive instability of prices lay in overall economic development; this was a gradual process and while economic strategies were being developed in vulnerable countries, there was a need for sizeable reserves.

(ii) National reserves could serve three purposes: emergency relief, stabilisation of prices, and providing "breathing space" to meet transitional shortages due to rapid economic development.

(iii) The costs of creating such reserves and storage facilities were beyond the capacity of most underdeveloped countries.

Our main conclusion was that such reserves could only be built by stepping-up aid on concessional terms.

It will be seen that the main preoccupation of the surplus-producing countries during this period of intense discussion was that international agencies should do nothing to interfere with the normal channels of trade. There was never any suggestion that surpluses could profitably be used for assisting the economic development of the less-developed countries. This came up in April 1959 when President Eisenhower urged Congress to adopt a "Food For Peace" Plan which was a bold attack on the problem of surpluses. Agriculture Secretary Ezra Benson was given charge of the Plan. In his letter to Benson, President Eisenhower referred to FAO's efforts to launch a worldwide campaign against hunger. Benson called a conference in Washington of the five major exporting countries — Canada, Australia, Argentina, France and the US. I was specially invited to take part on behalf of FAO. The Conference set up a Wheat Utilisation Committee which was to serve as a consultative body to the five governments and would maintain a close working relationship with FAO. The activities of the Committee were to include "making more effective the utilisation of wheat surpluses for the promotion of economic development, ... coordination of disposal programmes for economic development with other development activities, ... providing wheat to individual countries on concessional terms, safeguarding commercial marketings".

In Chapter 16, I referred to the steps taken by me to have the General Assembly Resolution drafted in a way that conformed to the ideas being developed under the FFH Campaign. I think I would be right to claim that on two essential points we were successful: (i) the Resolution endorsed the FFHC urging all members of UN and Specialised Agencies to support it in every way, and acknowledged the dominant part that FAO could play in dealing with this problem, and (ii) recognized the central theme that FAO was developing — that the ultimate solution to the problem of hunger lay in an effective acceleration of economic development of the underdeveloped countries using the large agricultural surpluses which could not be disposed of on commercial terms.

The creation of the World Food Programme was the final outcome of the General Assembly (GA) resolution which asked "FAO ... to establish ... procedures ... by which ... the largest practicable quantities of surplus food may be made available ... as a transitional measure against hunger, such procedures to be compatible with desirable agricultural development ... in the less developed countries." My report, *Development Through Food,* to ECOSOC formed the basis of the World Food Programme. In this report, I made the following points: as a first principle, surpluses in aid must be granted for the promotion of economic and social development. As a second basic principle the main decisions as to what course development should take in food-aided countries should be taken by the countries themselves, and not by the donor countries as had been the practice in the past. The third basic principle was that aid would be more effective if it was integrated in national development programmes, and became part of a national effort, in which the use of every resource should be related to every other resource for achieving nationally agreed objectives and aspirations. I made this observation:

> An operational plan for economic development ... is a charter of policy which has to outline what the people may expect in time, and also the duties they have to undertake ... It must receive sup-

The Three Circles Plan worked out by FAO was complete with procedures covering the period of warning, the period of action, and the post-emergency period. I had always held the view that in emergencies such as floods and famines which normally called for emergency relief, the international agency best qualified to take responsibility on behalf of the international community was FAO. In more recent years other organizations have been set up which overlap each other's efforts and do not have a proper mechanism for effective coordination. I suggested further study towards more flexible arrangements.

My report went through various stages of discussion — by ECOSOC, by the FAO Council and by its Committee of the Whole, and finally by the FAO Conference and the General Assembly. I need not reiterate the details but the picture would be incomplete if I did not acknowledge here the debt FAO and the United Nations owe to Senator George McGovern who came at the head of the US delegation to the Inter-Governmental Advisory Committee. It was here that he submitted a formal proposal for the establishment of the World Food Programme on a three-year experimental basis which aimed at raising $100 million in commodities and cash. This bold initiative by Senator McGovern, who same time to assist economic and social development. He also announced at that meeting that the US would pledge $40 million in commodities and cash. This bold initiative by Senator McGovern, who was then Director of the Food for Peace Programme in the White House, finally got my proposal off the ground. I name Senator McGovern personally because he had to contend with his two principal advisors, one from Agriculture and the other from the State Department who were against any definite commitment by the US. Senator McGovern is one of the finest, most liberal political leaders I have come across. He later contended for the Presidency of the United States and lost. I cherish him as a very close friend.

National Food Reserves and Post-harvest Losses

IT WAS GOOD to see that some of the old problems with which FAO had been grappling from the very beginning received special attention at the World Food Conference in 1974. I would like to mention two problems so that in the energetic steps now being taken, FAO may profit from the experience of the past. These are the building of national reserves and reduction of post-harvest losses.

In my first appearance before the UN General Assembly as Director-General, I addressed the Second Committee on World Food Reserves. The debate had been postponed to allow me an opportunity to make a statement. The debate centred on Document D/2855 — *Functions of the World Food Reserve: Scope and Limitations* which was a valuable analysis of World Food Reserves prepared by FAO. In my statement I referred to the unprecedented generosity of the United States in bilateral help (*Public Law No. 480*), and the large reservoir of surplus stocks now available in that country. I said that rather than wanting to compete with bilateral programmes or run them ourselves, it would be better for us to establish a cooperative relationship with the main bilateral donors and in this way make their bilateral assistance as effective as possible. Someone had used the phrase "coordinated bilateralism" in this context and this aptly described what I had in mind.

In my statement I also dealt with the problem of "surplus disposal" and referred to the new proposals made by the United States Delegation for the strengthening of national reserves in underdeveloped countries through the use of surplus food stocks made available on special terms. I supported the proposal because it shifted the emphasis from post-fact relief to preparedness before the event and also provided manoeuvrability in economic planning. I also stressed the need to keep in mind the principles of surplus disposal as worked out by FAO and accepted by member nations. This would avoid dangers of harmful interference with normal patterns of production and with international trade channels. The resolutions finally adopted by the Committee requested further study.

The debate showed real appreciation of FAO's work in this field and our continued active participation in any future studies was regarded as a necessity.

Some of the countries with surpluses were already taking steps to reduce their production to solve their surplus problem. While this was a perfectly understandable policy from a purely national angle it raised fundamental issues which were important to FAO. When more than half of the world was still going hungry it seemed a tragedy that problems of international marketing and distribution should cause the surplus countries to apply the old policy of restrictive production. The debate in the Second Committee brought the problem into full focus. While national policies had to solve national problems, we had to be careful not to go too far. We must not lose sight of the outstanding importance of ample stock-holdings of foodstuffs to provide reserves against unforeseen shortages.

Following the debate a good deal of work was taken in hand. An intergovernmental FAO working party met in Washington and studies were undertaken at Headquarters. I felt, however, that field enquiries were necessary to meet the challenge of the questions put to us. I felt that we needed to learn about the problems firsthand and that it was necessary to provide some realistic illustrations for the FAO study which would be presented to the Economic and Social Council. There was little time left for such field enquiries but thanks to the very helpful attitude of the

Governments of India and Pakistan it was possible to arrange for a small FAO staff-team to pay brief visits to both these countries. The conclusions of the FAO report with which the Economic and Social Council expressed agreement provided a most useful basis for a more thorough and practical approach to the problems involved.

The report prepared by FAO highlights the following factors. Firstly, while the main remedy against both chronic want and excessive instability of staple food supplies lay in overall economic development and higher productivity, this must be seen as a gradual process. In the meantime, there was need for sizeable reserves in vulnerable countries. Secondly, national reserves could serve three purposes: emergency relief against famine, stabilisation of prices, and "breathing space" to meet transitional shortages due to rapid economic development. National reserves should therefore be established on a multi-purpose basis to meet all three needs. Thirdly, the costs of creating adequate reserve stocks and storage facilities were substantial and beyond the capacity of most underdeveloped countries to finance.

Our main conclusions stressed the great difficulties for underdeveloped countries to find from their meagre resources the extra margin required for building-up reserves. The point was also made that notwithstanding all the special facilities which already existed for surplus disposal operations on concessional terms, real progress in regard to food reserves was not likely to be made unless further concessions were offered possibly by way of grant. Our report, despite its rather outspoken and fairly strong conclusions, was given a very good reception at the Economic and Social Council.

The Economic and Social Council recommended in its resolution that individual governments who wanted to establish or increase national reserves should prepare specific plans and invite governments, willing to assist in the establishment or enlargement of such reserves, to enter with them into discussions with a view to the early realization of mutually acceptable plans. After that the countries in need of national reserves could follow-up the resolution.

Coming now to the problem of post-harvest losses, I recall that the very first seminar I attended as newly elected Director-General was on the improvement of storage facilities for grain under tropical conditions. It was held in my home city, Calcutta. FAO had taken an early interest in this problem. Quantitative estimates of losses of grain in storage were difficult to assess and were too often underrated. A conservative estimate for the average loss would be as much as 25 per cent. In the 1964/65 Programme of Work we included a number of projects and activities on pest control, disease eradication in crops and animals, and improved handling and distribution of food. A number of surveys were carried out on the prevailing situation in dairy products, meat, cereals, fruits and vegetables, fish and other selected types of food. All this helped to bring the problem to the attention of Governments and the people directly concerned. We were also able to establish some small installations such as slaughter houses, processing factories for several kinds of food, and drying and storage facilities. In cooperation with the World Bank, we advised countries on the construction of larger scale food industries based on the latest scientific knowledge and experience. The great importance of the subject led the ECOSOC Advisory Committee on the Application of Science and Technology to study the outline of the problem we had presented. On the basis of this outline the Committee gave a detailed survey of the possibilities for reducing such losses. The possibility of "cold chains" to facilitate food preservation, distribution and marketing, not only inside one country but also between several countries, was one of the points we pursued further. Unfortunately, conditions today in many countries have not improved significantly.

The Green Revolution

IT WAS NOT UNTIL the middle of the eighteenth century that the sexual nature of plants was scientifically examined, and it was only at the beginning of this century that plant breeding for commercial agriculture came to be widely practised. With the rapid increase in knowledge of plant genetics and techniques of plant breeding, the 1960s ushered in an era of high-yielding seeds. The Green Revolution was a hopeful description for this new era. One of the prominent plant geneticists, Norman Borlaugh, won the Nobel Peace Prize but the honour as he himself graciously said was truly to be shared by the many who had worked to bring about the revolution.

FAO is not a research organization. FAO's function is to maintain the closest possible contacts with the leading national and international Research Centres and gives extensive publicity to promising lines of research through its periodical *AGRIS*. In its field operations FAO applies the latest techniques and knowledge under various conditions. These functions were pursued vigorously in my time and since, under successive Directors-General.

FAO Seed and Fertilizer Programme

In support of the Green Revolution, we took two important initiatives. The World Seed Campaign was launched in 1958 and was continued through 1962. The World Seed Year was observed in 1961. It was realized that the quickest, cheapest and most effective way to increase crop production and productivity would be to use quality seeds of adapted varieties suited to different ecological conditions, such as irrigated areas, rain-fed areas, marginal areas and areas prone to drought. The selection of indigenous varieties was encouraged. It was realized that mere supply of improved seeds was not enough; it had to be supported by trained manpower and the right institutions.

It soon became clear that for social and economic reasons the poorer farmers could not benefit from the Green Revolution. The whole situation was brought under review between 1968 and 1975 and, on the basis of those reviews, the Seed Improvement and Development Programme (SIDP) was established to reinforce and amplify the earlier efforts. The broad categories of activities of SIDP include: development of seed production and utilisation; provision of seed and planting material; training in seed production; distribution and supply of information; and training materials. The World Food Conference in 1974 in a Resolution recommended the creation of a Fund of $33 million through voluntary contributions to strengthen the SIDP. Though the target has yet to be reached, the resolution clearly endorsed the FAO Seed Programme.

Side by side with the Seed Programme, the FAO Fertilizer Programme had started in 1960 as a joint venture between the World Fertilizer Industry and FAO. The Programme grew from year to year. Starting with $300,000, the budget passed the $5-million mark in 1978, with 80 per cent coming from donor governments and 10 per cent from the Industry. The Programme operates in the following way:

(a) information is supplied on the kinds and amounts of fertilizer and related inputs needed for particular crops under different conditions;
(b) necessary extension services are provided for the efficient use of fertilizers and related inputs;
(c) attention is drawn to inadequate price policies and unfavourable relationships between crop prices and costs of fertilizers and other inputs; and
(d) credit and marketing systems affecting fertilizer use in the area of operation are kept under study.

A project is established in agreement with the recipient countries. The average duration of a project is six years. The work comprises simple trials and demonstrations on the farmer's own land, pilot schemes on fertilizer distribution and credit facilities, and intensive training courses at all levels. It is the training component which finds most support from the recipient governments and there is increasing interest in the creation of fertilizer and other inputs services. The MSA (Most Seriously Affected) countries are turning more and more to international agencies for financial assistance. The World Bank, the Regional Banks, and now IFAD* are making more credit available. The World Food Conference of 1974 recommended the setting-up of an International Fertilizer Supply scheme (IFS) by FAO to cope with crises arising from short

*International Fund for Agricultural Development.

supply and high prices. The IFS has since been created, and by 1979 it had supplied more than 500,000 metric tons of fertilizers to MSA countries.

However, to maximise production, distribution of high-yielding seeds and fertilizers was not by itself enough; there had to be a more comprehensive approach. Along with high-yielding seeds and fertilizers it was also necessary to have the right soil type, good farming methods and disease-control programmes. Provision of controlled irrigation facilities was another important factor. In pursuing our Seed and Fertilizer Campaigns, we tried to bring in the other components but naturally not always with success. The problem was discussed at the World Food Congress (1963) which made the following proposal:

> There is an urgent need to increase the production of food in developing countries as rapidly as possible. Fertilizers are a spearhead in increasing crop production. It is therefore proposed that a "pool" of production requisites, including fertilizers, pesticides, farm machinery etc., should be established under the administration of FAO in order to provide aid in kind to countries that require it.

This proposal for the creation of an Agricultural Inputs Pool received attention in several international discussions under the auspices of the FFHC. In November 1965, the FAO Conference took it up and agreed that the proposals for some form of international fund for fertilizers and other inputs merited careful study and asked me to undertake this work.

In his message to the US Congress on February 10 1966, President Johnson stated:

> We must encourage a truly international effort to combat hunger and modernise agriculture. We shall work to strengthen the Food and Agriculture Organization of the United Nations. The efforts of the multilateral lending agencies, and of the UN Development Programme should be expanded — particularly in food and agriculture. We are prepared to increase our participation in regional as well as worldwide multilateral efforts, wherever they provide efficient technical assistance and make real contributions to increasing the food-growing capacities of the developing nations.

Food Production Resources Programme

In the light of these developments, I formulated a proposal for a *Food Production Resources Programme* to be administered by the Director-General of FAO; the minimum target for such a programme during the first biennium was to be $50-million-worth of material which would consist of nitrogenous fertilizers, pesticides, farm machinery, etc; assistance to be provided as loans on a long-term interest-free basis so that the Programme would be in the nature of a revolving fund. I pointed out that FAO was directly concerned with the problem and was the most qualified to deal with it. The operations would mobilise a wide range of specialised knowledge possessed by the Organization. Moreover, through its worldwide network of field programmes, FAO could locate the countries and the areas where progress in agrarian reform and the improvement of marketing and credit institutions had created a favourable climate for the greater use of material inputs in agriculture. The exercise of this knowledge and experience would be an element in the supervision of the Programme. It would ensure that the fertilizers and other requisites made available by the donor countries were put to use under the most favourable conditions.

The years 1965 and 1966 saw a further decline in per capita production in the developing countries, and a sharp decrease in the North American grain stocks which had hitherto provided emergency supplies. This made it imperative for the developing countries to increase their own food production within the shortest possible time. I therefore modified my original proposal and called for a massive programme of *$500 million per annum* to create the necessary impact. Fully aware of the reluctance of the developed countries, I proposed that the programme should be in two parts, the major part — $450 million — to be in the form of bilateral aid. Since most of the developed countries produced some of these agricultural requisites, they could reorientate, without adverse effects on their balance of payments, their bilateral programmes accordingly. FAO could assist by indicating where the necessary conditions existed for effective schemes, by formulating realistic projects and making the information available to prospective donors and, where requested, by helping with the implementation and supervision of projects. The other part would be as originally suggested — a multilateral fund of $50 million a year. With this fund, FAO would also have the opportunity to undertake new types of programmes which when tested, could be adopted. These proposals came up before the 1967 Conference. It was left to the developed countries to decide for themselves what course they should follow. They were hesitant to assume responsibility for a new Fund though the yearly contribution was to be only $50 million on voluntary pledges and they were not in a mood to accept any obligation to channel their bilateral assistance in the way I had suggested.

The central argument of providing assistance to the developing countries to help them increase their agricultural production and reduce their dependence on imports was too potent to be set aside. The problem remained, to be faced another day. The idea surfaced again at the World Food Conference of 1974. This time, fortunately, it was not solely a matter for the developed countries — the Geneva Group as it is now known; the oil-producing countries — OPEC — with their newly found strength were prepared to meet the challenge. What evolved was, in some essential respects, quite different from my original proposals.

There was to be a new Specialised Agency of the United Nations, independent of FAO. In regard to identification and implementation of projects, it would form its own policies. But most important of all was the provision that the main objective would be not maximisation of agricultural production but assistance to countries most seriously affected (MSA) — the countries most in need. Clearly, this MSA approach was different from my spearhead approach. But there is a need for both and funds will have to be found for both. I believe the whole matter will come up for a review when the time for replenishment of the Fund becomes due and an assessment is made of the whole operation.

Chapter 21

FAO and the World of Commerce

The Industry Cooperative Programme

ONE OF MY major initiatives under the FFHC were the links I established with the major industries of the developed countries. I hoped this initiative would help to accelerate economic development in the developing countries of the world. I think I can best explain my approach by quoting the Letter I wrote to the Agriculture Ministers in August 1965:

> During the last few years FAO has changed from a primarily technical organization to become one of the world's most important development agencies, and is concentrating more and more on operational work. Out of some 65 million dollars to be spent this year through FAO, about 45 million dollars will be devoted to development operations.
>
> We have, moreover, deliberately moved in our field operations from studies and surveys to action and implementation. The agreement to set-up at FAO Headquarters a joint FAO/IBRD (International Bank for Reconstruction and Development, now called World Bank) division constitutes a major step in that direction. This arrangement is already beginning to yield tangible results and by the end of 1965 the work of the joint unit will have resulted in development loans and farm credits for very substantial amounts. A similar agreement was recently concluded with the Inter-American Development Bank.
>
> A further important step to speed up agricultural development is

the Indicative World Plan. The need for such a plan stems from the unsatisfactory rate of progress and the realization that we can meet the challenge only by setting quantitative targets and specific deadlines all of which need to be consistent with each other and governed by strict priorities.

In this global effort, a massive expansion of industries related to agricultural production and food distribution deserves a very high priority, especially since the possibilities in these fields are far from having been fully exploited. Local processing, easy and cheap transportation, and better storage are indispensable means to reduce waste which takes such a heavy toll of food output in the developing countries. Fertilizers, seeds, vaccines and pesticides are needed to step-up agricultural productivity — while balance of trade considerations make it imperative that more and more of these steeply rising agricultural inputs be produced by the developing nations themselves. FAO's particular concern and responsibility for the industries processing raw materials from farms, forests and the sea is determined moreover by the fact that they provide an important means for import savings and a potentially substantial source of export earnings. It needs to be realized that the hopes attached by newly independent nations to this type of industrialisation have so far not been fulfilled. Yet the obstacles hitherto encountered will continue as long as raw material production and industrial processing are not jointly planned and carried out in continuous and organic relation with each other.

Most processing industries are best located in, or closely adjacent to, rural areas and the same is true for the manufacture of farm machinery and other agricultural inputs. These industries are capable of providing vitally needed employment and incomes to rural people and will act as a brake on the exodus to urban slums.

For all these reasons FAO has always given considerable attention to the development of certain industries — both those devoted to the processing of farm, forest and marine products for which FAO carries direct responsibility, and those industries which help to raise productivity and improve distribution. We also have close ties with many industrial enterprises which specialise in establishing new processing industries in the developing regions.

In certain fields, like forest industries, pulp and paper manufacture, fish processing, and fertilizer application, continuous cooperation has already been established and has produced encouraging results. Usually this cooperation is achieved through advisory committees or special industry panels established under FAO's Regular Programme and more recently within the framework of FFHC. In addition, FAO's work has had the benefit of significant industry support for specific projects in fighting animal disease or in organizing research work and field demonstrations.

The United Nations and FAO owe a great debt to Senator George McGovern (right) for the establishment of the World Food Programme and the generous support given it by the USA

Planned by Darius the Great in the 6th Century BC, only a few carved pillars of the Audience Hall remain to testify to the grandeur that was once Persepolis

With Jaiprakash Narain of India (third from left) after his address to my staff in the FAO Headquarters

On numerous occasions, Pope Paul VI blessed the Food and Agriculture Organization for its work in helping the hungry people of the world

In 1964, the author (front row, third from left) was presented with an Honorary degree from Dublin University

Two Directors-General of the Food and Agriculture Organization

Prince Philip visiting FAO Headquarters in Rome

With Earl Attlee (left) and Lord Boyd-Orr (right) on the occasion of FAO's 20th Anniversary

The vast tropical rain forests of Brazil have an environmental importance for the whole of the South American continent and beyond (author, centre)

Under Charles de Gaulle's personal influence, generous measures were adopted to spread the message of the Freedom from Hunger Campaign

Eamon de Valera (front row, centre) was an hero of the Indian Revolutionary Movement; my wife and I visited him in 1964 while he was President of the Republic of Ireland

(Left to right): Author with U Thant, Secretary-General of the United Nations. He once told me that he was ve fond of watching boxing on television but he never favoured one side or the other

At the 1963 World Food Congress, President John F. Kennedy declared with earnest intensity that failure to conquer world hunger would be 'a disgrace to this generation' (author, third from left)

My wife and I with Queen Juliana of the Netherlands at her Soestdijk home

Left to right: Willie Brandt (then Mayor of West Berlin), the author, President Lübke (FRG) and Edgar Pisani (French Minister of Agriculture) during Green Week in Berlin

Among the ruins of Gerasa, Jordan (Mrs. Sen and author, second and fourth from left)

My last day in FAO (22nd February 1967), with Miss Jessica Campbell (right) and Mrs. Carmen Lobo-Checchi, my Personal Assistant and her Secretary

A confusing signpost – measured in hours, distances from Strömfjord Airport in Greenland

Walking through the Khyber Pass brought thoughts to me of India's monks who passed this way hundreds of years ago, travelling from the land of Lord Buddha to China

Some of my fellow travellers during my visit to Russia, across Siberia to Irkutsk and Lake Baykal

During my eleven years as Director-General of FAO, my travels brought me from the hot sands of the desert to the ice floes of the Arctic.

Opening of a Sea Fishing Centre at Pusan in South Korea

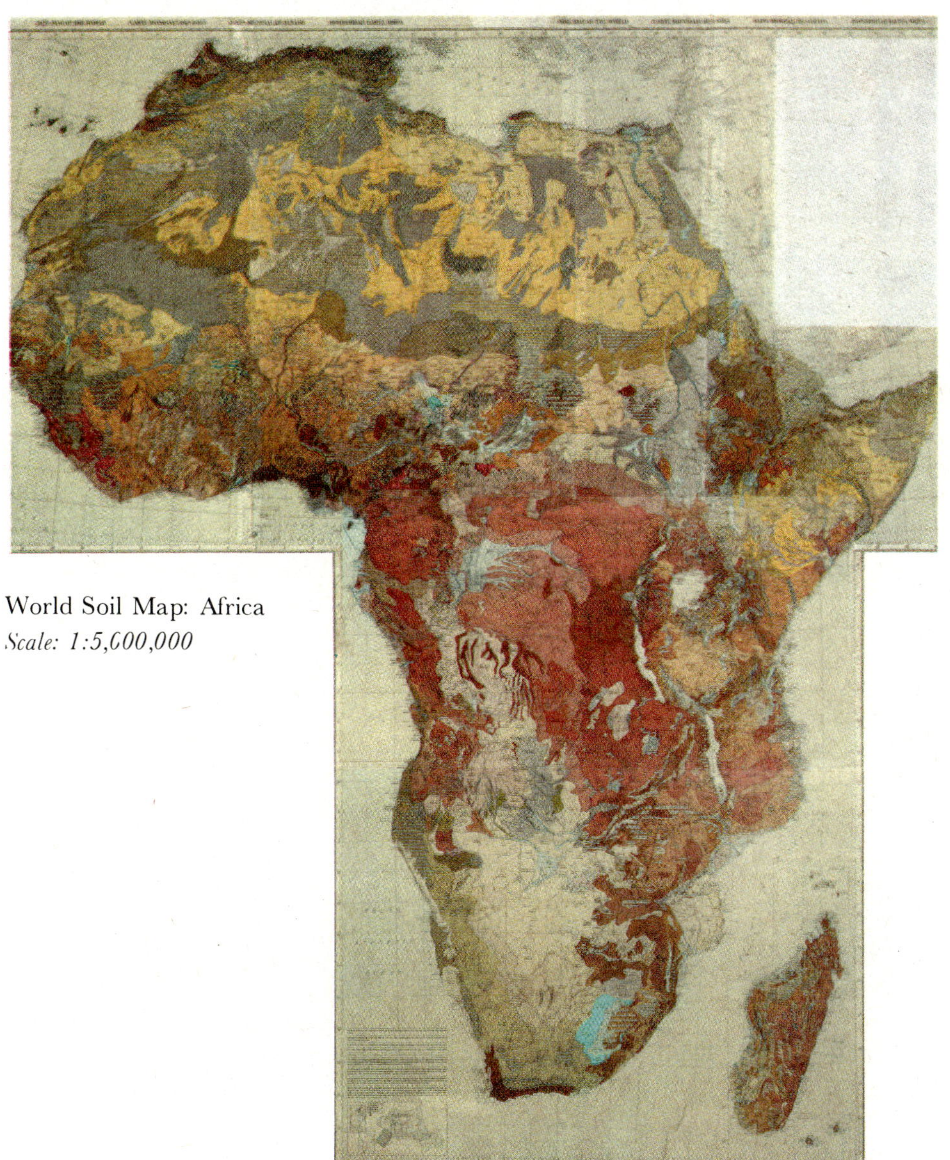

World Soil Map: Africa

Scale: 1:5,000,000

SOIL MAP
Soil Map - Bekaa Valley (Lebanon)
Legend:
1: Calcic Xerosols and Calcaric Regosols, medium to fine textured, undulating.
2: Calcaric Fluvisols and Gleysols, medium textured, flat.
3: Calcaric Cambisols and Rendzinas, fine textured, rolling to steep.
4: Chromic Cambisols and Lithosols, fine textured, stony, rolling.
5: Chromic Luvisols and Lithosols, fine textured, stony, rolling to steep.
6: Chromic Luvisols and Calcaric Cambisols, fine textured, flat to undulating.
7: Orthic Luvisols and Lithosols, fine textured, rolling.
8: Lithosols and Chromic Luvisols, fine textured, stony, steep.
9: Fluvisols and Gleysols, medium to fine textured, flat.
10: Fluvisols and Orthic Luvisols, fine textured, flat.
11: Fluvisols and Chromic Luvisols, fine textured, flat.
Note: Soil Association Boundaries (Soil Map of the World) outlined by thick lines :

anything else this shows the growing interest in agriculture in the developing countries for which both FAO and the World Bank can feel satisfaction. In 1976 and 1977 alone, 74 projects prepared with FAO's assistance were approved by the World Bank and by the other financial organizations which gave loans and credits of $1,545 million with total costs coming to $3,941 million.

The World Bank's link with FAO was a departure from its normal policy. The Bank had previously financed large irrigation projects which had benefited agriculture, but direct agricultural projects had not been undertaken. I had met George Woods when I was India's Ambassador in Washington in the early 1950s. As President of the World Bank he took part in the World Food Congress in Washington in 1963 which passed a resolution asking FAO to explore the possibilities of establishing an international source for financing the immediate production aids to small and medium-sized farms. Until then this had been regarded as a national rather than an international responsibility. However, now that people were more aware of international responsibility for promoting development in the developing countries, the question had to be tackled.

Early in 1964 George Woods said in a speech which was widely published that the Bank was ready to undertake a more positive policy towards agricultural problems than hitherto but it needed time to explore the various possibilities. As soon as this came to my notice I took a plane to New York to meet him and discuss his ideas. The result was the FAO/IBRD* Agreement. He elaborated his ideas at the Conference on Trade and Development in Geneva in March 1964. I remember having paid a tribute to him at the same session, for his imaginative leadership in proposing to expand the scope of the Bank's operations. Under the Agreement the two agencies proposed to cooperate in the identification, appraisal and end-use supervision of projects suitable for financing. While the identification and preparation of agricultural projects was to be considered primarily an FAO responsibility, the appraisal and supervision aspect would be the primary responsibility of the Bank. Provision was also to be made for technical assistance by FAO in the implementation of the projects. A Joint FAO/IBRD Division was created in FAO. Within a short time it proved an important arm of FAO in initiating development activities which before then were beyond our means. Later, the title of this Division was more appropriately changed to FAO Investment Centre, because the growing investments were undertaken under joint management. By 1978 projects worth $13 million were reported to have been undertaken and completed through the initiatives of this Centre. By far the greater part of the funds have come from the developing countries themselves. The Bank has in recent years also expanded its own direct activities in the agricultural field which

*International Bank for Reconstruction and Development, now called the World Bank.

now accounts for more than one-third of its annual expenditure in loans and grants. The imagination and forceful leadership of Mr. MacNamara, the Bank's President, has given a new dimension to its work in the field of agricultural development in the developing countries.

Chapter 22

FAO and Field Programmes

Funds — a Perennial Problem

DURING THE FIRST ten years of its existence FAO concerned itself largely with problems of post-war recovery. While a few bold ideas were developed they were never implemented because they were too far in advance of the mood of the times, for example, Boyd-Orr's proposal for a World Food Board. From the organizational and budgetary points of view this first period was somewhat static. The budget was kept at a fixed ceiling of $5 million a year from 1945 to 1950; between 1951 and 1956 it rose only to $6.6 million. When I came to FAO in 1956 it had crept up to $7 million. The Programme of Work itself followed a set pattern. One of my first acts was to try and break out of this straitjacket.

In presenting the first Programme of Work and Budget in 1959, I said that although FAO was at that time considered to be one of the foundation stones for post-war peace and prosperity, its budget had been kept frozen at too low a level. I asked if FAO could retain and develop its usefulness if its resources continued to fall far short of the needs that had to be met. I made the point that FAO had now completed eleven years of its life and that its effectiveness and efficiency as well as its potentialities had been subject to continuous review. The question now was whether governments wished to go forward or whether they wished to stabilise the activities of FAO at the current levels with the danger of eventual stagnation and decay.

Again, in presenting my Programme and Budget in 1961 I referred to the dynamic political and social changes in the world and the inescapable impact of such changes on the practical role of international organizations like FAO. I referred to the emergence of new nations, particularly in Africa, which had a very special need for guidance and help in almost every aspect of their social and economic development. In 1963, in the atmosphere created by the World Food Congress, I felt justified in proposing a Programme of Work and Budget that was much larger than anything previously proposed. I presented the picture shown up by FAO's Third World Food Survey and the declining trend of per-capita production, especially in Africa and Latin America. I quoted the inspiring words of President Kennedy in his address inaugurating the World Congress: "We have the ability ... we have the means, we have the capacity to eliminate hunger from the face of the earth in our lifetime. We need only the will ..."

The fight to increase the resources of FAO continued all through my time and has been pursued by my successors. I began with a biennial budget of $13.5 million which eventually climbed to $167 million.

The structure of FAO was based on the broad concept that it should be a centralised organization, and that it should maintain close contacts with the Member Countries and be aware of their problems through its Regional Offices and Country Representatives. In 1965 the FAO Conference decided to set up a high-powered Review Team to explore the organizational structure that could best contribute to the achievement of FAO's goals "in the most effective, economic and appropriate manner". The main reason behind this move was the disquiet caused by the rising budgets for which I had been campaigning. The Review Team completed its work in a year and prefaced its report with the observation that "FAO's critical administrative problems have been accentuated by its movement into a major new activity of operation, on a scale that has brought about a fourfold budget increase during a period of only seven years."

The Review Team proposed ideas for "radical restructuring and reorganization" on the grounds that "the existing structure and organization were basically developed for an entirely different programme — data gathering, worldwide and area meetings and publications — and have been seriously outgrown by the shift towards developmental work in the field." The main burden of "radical" recommendations was to divest the Technical Divisions at Headquarters of all their present operational responsibilities, confining them to study, research and technical meetings as in the past, and to concentrate all operational policy and management in four Regional Offices to be created at Headquarters by a transfer of those offices from the regions. I had to take a stand on this proposal. In presenting my report to the FAO Conference I pointed out that "operations" and "expertise" had special meanings in FAO. FAO's expertise consisted not only of up-to-date knowledge but also of experience in the application of that knowledge in the field. Ever since FAO began its field operations, its technical experts were fully involved. The insight and experience thus gained in problems of development in all parts of the world augmented and reinforced the academic and scientific qualifications which the experts brought to the Organization. If the technical experts at Headquarters were to become merely a staff to advise the field operators, the unique value of FAO expertise would largely be eroded.

In the summer of 1965 my wife and I were having a short vacation in the mountains of Spain. After looking at the *Siete Picos* (Seven Peaks) for several days, I got restless and wanted to get the framework for the Review Team off my mind. It took me nearly a week. I had no FAO contacts and no reference papers. The note emerged naturally as a *tour d'horizon.* On my return to Rome I wondered if this "Spanish Note" (as it later came to be known) should form the basis of a document of greater depth and detail. I finally decided to leave it as it was. I remember when this Spanish Note was circulated people were rather impressed that so much work was achieved "sitting on a mountain top". The Conference itself accepted it as "the comprehensive framework presented by the Director-General".

The Conference had an inconclusive debate on both the reports, leaving things more or less in the hands of the Director-General. I will only add here that I had another reason for strengthening FAO's Regional Offices. The growing importance of the UN Regional Commissions, whose terms of reference in social and economic matters were almost as broad as those of the United Nations Secretariat in New York, also called for a strong parallel FAO presence so as to avoid having people making any serious incursions into our fields of work without having our background and expertise.

The intimate relationship between work at Headquarters and in the field should be clear from what I have said. The object of strengthening the budget every biennum was intended primarily to strengthen our operations at country level. The main lines of field activities that are now followed by FAO mostly originated in my time under the thrust of the increased budgets. The main categories of field projects may be summarised as follows:

(a) Investment studies to assist the preparation of projects for investment, including surveys of agricultural, forestry and fishery resources.
(b) Development and management of resources essentially giving support to national services on land and water management, crop or animal production and development or exploitation of forestry and fishery resources during planning and implementation phases.
(c) Strengthening of National Institutions by advisory and supporting services.
(d) Training of personnel, usually in-service, with full involvement of a local institution.

(e) Area development by multi-disciplinary projects involving planning and/or implementation in specific geographic areas.
(f) Miscellaneous activities such as agricultural planning, food processing, nutritional programmes, regional and national workshops, expert consultations, fellowships (André Meyer), supply of equipment and so on.

The efforts made by FAO over the years now manifest themselves in various ways. The growing self-reliance in many developing countries who prefer to use their national rather than FAO expatriate experts is one example. Another is the increasing importance of national institutions. FAO now employs nearly 1500 professional staff from 85 countries in various field programmes. Geographically, 34 per cent of the total funds for FAO/UNDP field projects, up to the end of 1977, went to Africa, 24 per cent to Asia, 20 per cent to Latin America, and 20 per cent to the Middle East, North Africa and Europe.

I have already described the work of the FAO Investment Centre which since its inception in my time has been responsible for projects worth $13,000 million.[1] The other two important sources of funds for FAO field programmes have been the UN Development Programme (UNDP) and Trust Funds provided by individual countries.

FAO has always been the largest recipient of UN Expanded Technical Assistance Programme which later developed into UNDP. I have taken an active part all through the years in many discussions on the administration of the Funds, both as India's delegate to ECOSOC and also in my capacity as FAO's Director-General.[2]

In the peak years FAO spent nearly a third of UNDP allocations in field projects (about $122 million in 1975). Up to 1972, countries often requested UNDP assistance on a project-by-project basis. Since that year, each country is allocated an Indicative Planning Figure (IPF) projecting the dollar value of UNDP assistance that the country might expect to receive over a five-year period. The country then submits to the UNDP Governing Council a total package of coordinated projects, closely related to national priorities, and integrated with other assistance efforts. Agencies such as FAO are then selected by UNDP in consultation with the Governments concerned to carry out the projects. This certainly was a most imaginative initiative which clearly recognized the desire of the Member Countries to be more and more self-reliant in drawing-up their national development plans.

Unfortunately, some differences have arisen over the old question of relationship between the UNDP Resident Representative in the country and the Country Representatives of the Agencies. FAO Representatives, having the largest share of field operations, would naturally

[1]See *The World Bank*, in Chapter 21, page 214.
[2]One of the controversies I was involved in is related in *UN and FAO – Some Recollections*, in Chapter 23, page 251.

be involved in these difficulties. The essential point to remember is that "coordination" should not impair the initiative of the Specialised Agencies. UNDP is a funding organization and its Resident Representatives are seldom experts in any field except in general administration. The "Consensus" arrived at in 1971 seems to me still to be valid. Meanwhile, FAO's persistent defence of its legitimate position appears at times to have been a reason why UNDP diverted some of its projects to other Agencies or even gave them to outside bodies so that the traditional FAO share of UNDP allocations has tended to decline. I will discuss, under the following section *Increasing Agricultural Production: The Scale of Investment Required,* the urgent need for all concerned to concentrate on raising agricultural production looking towards the year 2000.

The institution of Trust Funds received a great impetus under the FFHC. Several developed countries of Europe came forward to establish cooperative arrangements with FAO to carry out some of their projects in the developing countries for which they needed FAO's guidance and expertise. Trust funds were provided both by governments and non-governmental donors. Sweden remained the largest government donor in 1977, followed by Denmark, Belgium, Norway, Canada and Switzerland. The Netherlands, Finland and USA also participated. While Government Trust Funds amounted to $24 million in 1977, non-governmental donors put up $1.6 million. These Trust Funds accounted for nearly half of the extra-budgetary funds spent by FAO in 1976 and 1977 on field programmes. A feature nowadays is that developing countries are coming forward to channel through FAO the money needed to finance some of their own national development programmes. These developments clearly demonstrate the confidence that FAO has been able to create in its expertise and management capacity, both among the developing and developed countries.

Increasing Agricultural Production: The Scale of Investment Required

FROM TIME TO TIME attempts have been made to compute the level of investment necessary, both nationally and from external aid, to step up agricultural production in the developing countries. In my report *Development Through Food* which formed the basis of the World Food Programme, I quoted the figure of $10 billion a year which was the estimate of an Expert Committee set up by the UN in 1951. While working on my report I had been authorised by the FAO Council to call a committee of experts to assist me. This committee had come to a figure of $5 to 6 billion. All these exercises were, however, rather theoretical in nature. The first real attempt made by us was while preparing for the

Indicative World Plan for Agricultural Development (IWP) discussed in Chapter 18 (page 182).

The study made a comprehensive assessment breaking it into subsectors by capital and recurring expenditure. The effective demand for food was estimated to rise at 3.9 per cent per year. At the existing production rate of 2.7 per cent per year the supply gap would have to be met by imports costing $40 billion at 1962 prices. To meet this deficit, production would have to be raised to at least 3.7 per cent per year. This would require a capital investment of $5 billion a year in internal and external resources. Regarding recurrent inputs (fertilizers, pesticides, fuel and maintenance costs of farm machinery) the estimate was that the expenditures would increase from $6 billion in 1962 to $26 billion in 1985, an addition of $0.8 billion a year.

These figures were updated for the World Food Conference of 1974. The total investment requirement was estimated at $11 billion a year. This, however, excluded post-harvest technology, machinery and equipment, livestock inventory, agro-industries and institutional services. If all these were added together, the annual investment would rise to $20 billion. This would mean doubling the annual investment from that estimated in the IWP. Without this acceleration, the cereals gap in 1985 was likely to increase to 85 million tons requiring $14 to 16 billion in foreign exchange. The 1972 annual flow of external assistance for food needed to be increased from $1.5 billion to $5.3 billion, indicating a sharp increase from 8 to 10 per cent of the total investment to over 33 per cent.

The World Bank in a study in 1977 attempted to determine the resource requirements, both internal and external, for meeting the food needs in 1985 of thirty-five of the "poorest countries" with per-capita income of less than $200 in 1975. The food-grain deficit of these countries in 1985 was estimated at 45 million tons after allowing for production at current rates, and also allowing for likely commercial import and nutritional requirements. To bridge this gap, production must grow at double the present rate of 2 per cent per annum. This would mean an annual investment requirement of $4 to 6 billion at 1975 prices — a doubling of investment over a five-year period. In terms of external resource requirements, concessional assistance would also need to be doubled to $5 billion a year in the period 1976-81.

The International Food Policy Research Institute (IFPRI) study of 1979 covers thirty-six developing countries with a per-capita income of less than $300 a year. The study estimates the food deficit on current trends of production at 56 to 74 million tons by 1990, requiring an average production growth of 4 per cent per year to eliminate the deficit as against under 3 per cent at present. The total capital requirements to raise food production by 90 million tons, beyond the growth projections of 80 million tons, in order to overcome deficits, are estimated at $95 billion in the period 1975-90 at 1975 prices. Of this $57 billion would be needed in the first ten years or $5.7 billion annually. Additional capital would be needed for irrigation, land development, technology and infrastructure, the estimated cost of which would be $2.5 billion a year. In simple terms, all this would involve a doubling of present efforts. This is broadly in agreement with the other two estimates mentioned.

Finally, FAO's latest long-term perspective study *Agriculture: Toward 2000,* which is its main contribution to the preparation of the new International Development Strategy for the 1980s, makes estimates of the investment and external assistance requirements for achieving an annual average increase of around 4 per cent in the agricultural production of the developing countries during the 1980s. It is estimated that annual investment requirements at 1975 prices would rise, on the basis of "modest growth", to $31 billion net of depreciation by 1990, and $51 billion gross. This is on the basis of the OECD's narrow definition of agriculture, which excludes rural development and infrastructure, agro-industries, fertilizer production, and regional and river basin projects. Annual requirements for recurrent expenditures (for such things as fertilizers and other inputs) are estimated to be almost as large as those for investment.

Commitments of external assistance to agriculture (again using the narrow definition) are estimated to have reached $5.1 billion at 1975 prices in 1978. Although this is almost double the level in 1973 (the first year for which comparable data are available) it amounts to only about 60 per cent of the estimated requirements of $8.3 billion noted in the Manila Declaration of the World Food Council in 1976. The figure of $8.3 billion represents the value at 1975 prices of the $5 billion included in the *Guidelines for International Agricultural Adjustment* which were adopted by the FAO Conference in 1975. FAO's latest estimate is that these requirements would rise to a minimum of $12.5 billion a year by 1990, of which about $10.2 billion would be for capital investment.

Each of these five studies has a different approach and methodology. Their time frames and coverages vary. Yet, the basic conclusion from all of them is that food production in the developing countries must be increased very substantially from the present rate of about 3 per cent a year. These studies also suggest that to obtain the annual growth rate of 4 per cent, the annual investment would need to be doubled over the present levels. In terms of external resources, the World Food Conference of 1974 estimated an annual requirement of $5.3 billion at 1972 prices. This too represented a doubling of the on-going flow. The Conference estimate of external assistance of $5.3 billion included both concessional assistance and other official flows. The World Bank estimate of $5 billion at 1975 prices referred wholly to concessional aid.

The urgent need to concentrate on raising agricultural production very substantially both by national efforts and by external aid is thus abundantly clear. At present only about 17 per cent of external assistance is going to agriculture.

World Soil Map

THE IDEA OF preparing a Soil Map of the World was first conceived at the Sixth International Congress of Soil Science in Paris in 1956. The purpose was primarily scientific — to establish a common legend to overcome the lack of correlation.* It was also realized that such a Soil Map would be useful to the newly independent countries in developing their national plans for agricultural activities. But this was a secondary consideration.

When the project was brought to my notice I had already embarked on the idea of a Freedom from Hunger Campaign and I saw the possibilities that a World Soil Map with correlated legend could have in extending areas of agricultural production especially in Africa. My first opportunity to stress this aspect came in 1960 at the Seventh International Congress of Soil Science (Madison, USA). The Resolution included these words: "Considering the great need for soil maps that will integrate knowledge of soils of the world as a cultural and educational media designed to promote freedom from hunger ...". The Congress asked the Director-General of FAO in cooperation with UNESCO and others concerned to prepare the Map. In 1961 Dr. Veronese, Director-General of UNESCO, and I signed an agreement for joint preparation of the World Soil Map at scale 1:5,000,000. Following this Agreement, I approached the United States Government for their cooperation and I was told that the Madison Congress had not asked FAO/UNESCO to prepare an integrated soil map of the world, or any new map, or to attempt to develop a uniform system of soil classification and nomenclature; the decision was one of FAO and UNESCO. The scale adopted was also too large — 80 miles to the inch, which only the most skillful and experienced soil scientists, using supplementary maps of geology and relief, could read usefully. They offered their cooperation only to the extent of a soil map for the United States on our scale (1:5,000,000) warning us to consider carefully before proceeding with the project. We considered and we proceeded. In 1964 the first draft international legend was presented to the Eighth International Congress of Soil Science at Bucharest. By 1965 the global coverage had been worked out.

*"This work was undertaken primarily", write Drs. Dudal and Batisse of FAO and UNESCO, "in an attempt to overcome the lack of correlation that has hitherto characterised the field of soil cartography, in which various countries had established their own nomenclatures, survey methods and systems of classification. In particular, those who favoured a genetic classification of soils found themselves in opposition to those who advocated a classigication based on morphological criteria. This made any comparison or exchange of information extremely difficult. The point had been reached at which the nomenclatures used could not be translated because terms which at first glance appeared to have equivalents in other languages did not, in fact, cover the same realities; this entailed a very great risk of misunderstanding and practical error."

The publication was completed in 1978 after seventeen years of patient and sustained work. The Soil Map of the World is composed of 18 colour map sheets (format 76 x 110 cm, scale 1:5,000,000), which can be assembled as a unit, and also includes a legend sheet. The map sheets are grouped under major regions each of which is described in a separate explanatory volume.

I feel happy that this great cooperative international enterprise has borne fruit. Already, steps are being taken to use the Map as a basis for global assessment of land-use potential, soil degradation and desertification. The Food Plan for Africa discussed at the African Regional Conference in September 1978 drew much useful material from the World Soil Map. It was possible to envisage an additional 31 million hectares being brought under cultivation by 1990 if the required investments were available. The Food Plan for Africa offers an overall blueprint for restoring self-sufficiency in food to 94 per cent by 1985 and maintaining this level up to 1990. These are very broad conclusions and probably require much more detailed in-depth work. I would like to pay tribute here to those colleagues of mine whose faith and patience made it possible to complete this task, particularly Drs. Bramao and Dudal in FAO and those who cooperated with them in UNESCO.

Land Resources for the World's Food Production

AT THE 12TH International Congress of Soil Science, held in New Delhi in 1982, the question of Land Resources for the World's Food Production was discussed.[1] It is estimated that the world's population could stabilize at 10.5 billion people in the year 2110 — nearly two and a half times greater than the present 4.4 billion. To sustain the world's 6 billion people in the year 2000, agricultural production will need to be increased by 60 per cent during the next two decades, with a further doubling of food production foreseen in the next century.

The question arises — is there enough land to feed these future generations? The results of the FAO/UNESCO Soil Map of the World[2] provide a global answer: if the people of the world were to live in harmony, if resources were shared, if all cultivable land were used in an optimal way and if there were unrestricted movement of produce, then there would be food for all for many years to come. Sadly however, the reality is different: land resources are unevenly distributed between and within regions; technological inputs are in limited supply in developing

[1]The following information has been compiled by R. Dudal, G. M. Higgins and A. H. Kassam of the Land and Water Development Division, FAO, Rome, 1982.

[2]FAO (1971-1981) *FAO/UNESCO Soil Map of the World, 1:5,000,000*. Vols. 1-10, UNESCO, Paris.

countries; the gap between rich and poor is steadily widening; and the movement of food from surplus to deficit areas is hampered by difficult communications and unfavourable balances of payment. It is not enough for the world as a whole to have the capability of feeding itself; it is necessary to produce more food where it is most needed.

If developing countries are to reach a greater degree of self-sufficiency in their food production, available land resources must be matched with the needs of present and future populations. Only when land potential is quantified, in terms of potential population-supporting capacity, can the attainable degree of self-sufficiency be realistically assessed. In a recent FAO/UNFPA study,* a land inventory based on the World Soil Map and climatic data was used to calculate calorie-protein production potentials and, hence, potential population-supporting capacities, under various input, crop mix and conservation assumptions.

The crops considered in this study are among the most widely grown crops in the world — pearl millet, sorghum, maize, rice, wheat, soybean, *Phaseolus* bean, sweet potato, white potato, cassava, groundnut, banana/plantain, sugar cane, oil palm and grassland/livestock. Each data set of the land inventory was analysed separately for all crops and livestock production, to ascertain which use is the most productive under the unique circumstances of the areas' soil and climatic conditions. Prior to this analysis, deductions were made for land required for non-agricultural use, irrigation and rest-period (fallow) requirements. Limitations imposed by degradation hazards and the present mixture of crops grown, according to levels of inputs, were also taken into account.

The world, as a whole, has enough land to produce food for present and future populations. However, with the uneven distribution of land resources, population and agricultural inputs, food production falls short of requirement in many countries. In order to avert dependency on supplies from outside, these countries will need to increase their domestic food production.

When planning for a higher degree of self-sufficiency, it is essential that differences in land-resource endowment and crop-production potentials be fully appreciated. In some countries, land reserves are such that cultivation can be expanded to meet national needs and even beyond. In other areas, however, the limits of cultivable land have already been, or are about to be, reached and most of the increased production will have to come from the intensification of agriculture on land presently cultivated. Certain countries, with unfavourable soil and climatic conditions, may not have the means to meet the food requirements of their people, even if the level of inputs were to be optimized. Furthermore, other needs must also be met, such as fibre for clothing, raw materials for housing and industry, lumber and fuelwood, environ-

*FAO (1980) *Report of the Second FAO/UNFPA Expert Consultation on Land Resources for Populations of the Future.* FAO, Rome.

mental conservation and possibly export crops. Therefore, a balance should be established within each country, matching needs with the suitability of the land base for the various types of use.

The precarious food situation in a number of developing countries indicates that the mere availability of land is not sufficient to fill the gap between supply and demand. Incentives need to be created for farmers to remain on the land and to make it more productive. Priority must be given to rural development in terms of investment, pricing policies, energy allocation, access to inputs, technology transfer, transport, credit, training and research.

With the identification of critical areas in various parts of the world, it is apparent that future needs will have to be assured by a global food system that establishes complementarity of production between areas of different suitability. This interdependence implies that both national and global resource assessments are needed and that the responsibility for conserving and improving the common heritage of the world's lands should be shared by the entire international community. The World Soil Charter was adopted in November 1981 by the 21st Session of the FAO Conference. The Charter established a set of principles for the optimum use of the world's land resources, for the improvement of their productivity and for their conservation for future generations. It calls for a commitment on the part of governments, international organizations and land users in general to manage the land for long-term advantage rather than for short-term expediency. This Charter provides a framework for good stewardship of the resource base on which the survival of mankind depends.

Remote Sensing

ONE IMPORTANT development which has come into prominence since my experience in map-making is Remote Sensing.

Remote sensing is an expression that was first used in 1958 to describe airborne techniques designed to enable man to gather information about relatively distant objects on or near the Earth's surface. It includes aerial photography, which is perhaps the best known and most widely used of these techniques, and thermal scanning, side-looking radar and remote sensing by satellites.

Today, many organizations throughout the world are realizing increasingly the potential of remote sensing techniques to provide timely and cost-effective answers to problems in exploration, development, planning or management of renewable and non-renewable resources and environmental conditions. This has been particularly stimulated by the widespread availability of imagery from the Landsats and the environmental satellites, including NOAA and Meteosat, and this is leading to increased use of multi-stage sampling involving aircraft and other data-gathering techniques.

Because of the obvious economic and strategic benefits from remote sensing there is a growing requirement in developed and developing countries for centres of expertise to be set up to advise users on applications to disseminate data and to support field studies. This requires development of special facilities and training of personnel, since such facilities or personnel are often not locally available.

At FAO, remote sensing began with aerial photography over 25 years ago, in its early field projects. At the outset, of course, ordinary black-and-white (panchromatic) photographs were used; but the introduction in the 1960s of reliable colour photography and, more recently, infrared colour photography has led to new applications in resource surveys. At the present time, geometrically corrected aerial photographs (known as orthophotographs) and assemblages of these (known as orthophoto-maps) are being used directly in place of maps. Also, photography is employed extensively for surveys of land form, geology, land use, vegetation, forest types, forest strata and soils; high-altitude colour infrared photography has been developed for national coverage related to land use and soil survey (for example, in Sierra Leone).

As aerial photographic techniques have been refined and become more sophisticated, skill in this medium has taken on ever-increasing importance. Aerial photography has been or is being used in FAO-executed projects in countries throughout the world — for forest inventories in Argentina, Bhutan, Cameroon, Gabon, Malaysia, Peru and Senegal, to name only a few; for an agro-industrial development project in Bolivia; for an irrigation project in Dahomey; for a land capability survey in Indonesia; for river-course mapping in the Republic of Korea; for a land-use study in Morocco; and for soil mapping in Ethiopia, Nepal, the Sudan, the Syrian Arab Republic and Thailand.

Forestry

UNDER THE CHARTER, Forestry is FAO's responsibility in the UN system because it is primarily a question of land use, between agriculture and forestry.

Forestry is a subject which always had a romantic as well as a practical interest for me. In ancient Indian literature it was to the forests that the holy men Rishis and Munis used to retire after their earthly responsibilities were discharged to find communion with God. It was in the forests that some of them founded their monastic schools, an event which is portrayed in one of our classic poems — *Shakuntala* by

Kalidasa (5th Century). I have kept faith with this cultural background despite my exposure to Western education.

Towards the end of October 1956, on my way to Rome to assume charge of my new office, I stopped off at Seam Reap in Cambodia (now Kampuchea) to see Angkor Wat, the city with the magnificent ruins of an advanced civilisation engulfed by forests and lost to human memory for centuries. I saw what forests could do to obliterate human civilisation. I specially remember the beautiful edifices in the coils of the long creeping roots of the rain trees rather like a hapless prey in the coils of a boa constrictor! The Maya ruins in Mexico had also been engulfed by the forest for centuries.

In my first long official trip to Latin America, I journeyed into the interior of the Amazon forests. And in 1960 when I went to attend the Fifth World Forestry Congress at Seattle I saw the redwood forests, where some trees had stood for two thousand years.

I had a practical interest because as a young man I was a member of a Forestry Regeneration Commission in my part of the country and became initiated into the various aspects of Forestry problems. I had read accounts of the denudation of forests in the Himalayan region — there is an account by a British officer involved in the Survey of India 100 years ago who described a rich primeval forest in the upper reaches of the Ganges which no longer exists. The recurring and destructive floods in India which become more devastating as the forests become denuded had impressed on me the importance of forestry for the preservation of soils and for regulating the flow of river waters.

It is to be remembered here that until the end of the last century forestry was regarded mainly as a national asset to be "preserved" for climatic and other reasons (one being the control of flow of water in the rivers). Scientific "utilisation" of forest products came later under the impact of industrialisation in the Western countries.

I made it a point to attend and address both of the World Forestry Congresses held in my time — the Fifth at Seattle (1960) on *Multiple use of forest lands* and the Sixth at Madrid (1966) on *The role of forestry in the changing world economy*. My main theme at these Congresses was the important role that Forestry had to play in the whole panorama of economic development.

One of the principal studies done in my time was the *Timber Trends and Prospects in the Asia-Pacific Region,* with the cooperation of the UN Economic Commission for Asia and the Far East. This study revealed a rather alarming situation. The level of wood consumption in the region was extremely low, and, on the basis of very conservative assumptions regarding economic growth, needs of industrial wood were expected to double by 1975. Even at the present low level of consumption, however, the region had an adverse trade balance in forest products of some 225 million dollars, corresponding to 3.5 million cubic meters of roundwood equivalent. Taking the most optimistic view of the existing plans for expanding forest production, the region had to face a deficit of over 20 million cubic meters of industrial wood by 1975. This gap would imply a future outlay in foreign exchange of somewhere between 1.5 and 2 billion dollars if essential needs — of constructional timber, of cultural and industrial papers — were to be met. Ministers and Heads of Forest Departments of the region meeting in Delhi to discuss the Report agreed on the need for a drastic reappraisal of plans for forest and forest-industry development. The need was to plan forest development and future forest output with expected future requirements firmly in mind — as was already being done wiuh conspicuous success in one or two other countries of the region. For the forester the task was simplified by the possibility of creating new forests of quick-growing, high-yield species, while on the industrial side a careful integration of forest-industry expansion into general economic development plans should bring about an optimum utilisation of indigenous resources and savings in imports. Alas, how little progress has been made to date!

Another initiative I remember was the fairly detailed plan we prepared at the request of Turkey for the conservation and development of forests in that country. This plan was one of the results of the Antalya development scheme which was a major study under the Mediterranean Development Project, spanning the years 1957 to 1964 (already referred to in Chapter 18, page 181).

During my time, the drawing up of a *Convention on the Conservation of Nature and Natural Resources* for Africa was entrusted to FAO and UNESCO by the Organization of African Unity. This interested me particularly from the point of view of wildlife and conservation. As far back as 1936 I remember having drawn up a Convention for north-east India which I circulated to the authorities. The joint FAO and UNESCO Convention was formally adopted by the Organization of African Unity in September 1968.

The harmonious and intimate relationship that Nature maintains between forests, rivers and mountains is never quite realized until that harmony is disturbed and disasters follow. Apart from making for stabilisation of climate and weather, forests regulate the water flow that feeds the rivers. If forests are destroyed or denuded the waters flow out into the rivers unchecked often causing heavy and destructive floods. The denudation of forests also causes soil erosion, silting up of rivers, raising the bed level and thus forming another major cause for destructive inundation in times of concentrated rainfall. Nowhere has this been more apparent than in the Indian subcontinent. This is a matter of the utmost importance, involving four of the most populous countries of the world — China (because two of the mightiest rivers in the world — the Brahmaputra and the Indus rise in or near Lake Manasarovar in Tibet), India, Bangladesh and Pakistan. The situation calls for urgent measures by the countries of the Indian subcontinent as well as by the International Community. The World Bank will have a major part to play.

Fisheries

THE ROLE OF FAO in Fisheries development, as in other fields, is governed by the basic aims of the Organization, namely, improving the nutrition and well-being of rural populations and the promotion of economic development. In fisheries, this role assumed special significance in my time. In the more advanced countries, the introduction of larger modern ships with new types of gear and electronic aids permitted the extension of fishing operations into ocean areas not formerly exploited. Modern processing and marketing similarly changed the patterns of domestic and international trade. Large-scale commercial enterprises were generated alongside the fishing industry. There was an equally impressive development of government services and institutions, in particular, governmental and industrial research to accommodate long-range oceanic investigation, and modern engineering and food technology were applied to fish production and trade. In the developing countries, the continuing shortage of animal protein — mainly in the form of shortage of fish for the growing population — directed government attention more and more to the potentialities of the seas and inland waters. In many of these countries, both the gradual transformation of small-scale backward fishery operations into larger modern commercial enterprises, and the creation of entirely new fishing industries, were attempted with varying degrees of success.

FAO, being the leading intergovernmental body involved in these growing problems, had to take a fresh look at its own role. I proposed a strengthening of the Fisheries Division in every one of the biennial budgets. But the increases allowed were only on a modest scale, not really in keeping with the rapidly growing demands on FAO. The 1961 Conference approved my proposal for an Advisory Committee on Marine Resources Research and a Panel of Fishery Experts. The 1963 Conference agreed to create the nucleus of staff necessary to work on stock assessment and to strengthen the inland fishery resources and boat sections, and to our work in Africa. At the same time, the Conference expressed "concern at the inadequacy of the staff and funds in a field which was so essential for supplying the world with high-quality protein foods". It also raised the question as to how fishery activities could be given full recognition "in the Organization and among other interested bodies ...". Following the Conference Resolution, the Technical Committee examined the whole position further and observed *inter alia*, "The FAO through its Fisheries Division should be coordinating and guiding these developments rather than standing by and watching them develop and progress, but the Division finds itself unable to handle the task." This enabled me to make a case for raising the Division to the status of a Department with certain strengthenings all along the line, and my Programme of Work and Budget for 1966/67 was framed accordingly. In all these discussions, the Head of the Division, Dr. Roy Jackson, who had come to FAO with a background of experience in the International North Pacific Fisheries Commission, played a significant part.

However, these developments took place in a situation where, except for the territorial waters of 6 to 12 miles, the sea was open to all who had the capacity to fish, which meant mainly the North American and European countries, the USSR and Japan. Very soon, the UN Convention on Fishing and Conservation of the Living Resources of the Sea (1958) and the Resolution of the Law of the Sea Conference (1960), which had dealt with the *status quo,* were overtaken by dramatic developments arising from the demand of coastal states for an "economic zone" along their coastlines in which they should have the exclusive right to exploit the resources. (In the 1960 UN Conference it appears that the question of "fishery limits" was tentatively raised; the question was first raised on a regional basis by Peru, Ecuador and Chile over twenty years ago.) The matter came up with great vigour, I remember, at a FAO Council debate in 1966 in the form of an "Exclusive Economic Zone" (EEZ as it is now called) of 200 miles from the coast. This has since been adopted by the United Nations and forms the basis of the various actions being undertaken by FAO and UN to give effect to it, specially by extending assistance, technical and otherwise, to the developing coastal countries.

The first fishing industry to become highly mechanised and to operate with large vessels at long range was whaling, especially in the Antarctic. A great expansion took place in the late 1920s and by the mid-1930s resource depletion was evident. The International Whaling Commission (which was formed in 1946 to bring order into this industry and to conserve whales) failed in its attempts to do either. In 1959 FAO was asked to help in making an independent review of the scientific evidence. On the basis of this review the International Whaling Commission drew up some new regulations. Three years later, when some whaling countries tried to back out of their commitment to these new regulations, I found it necessary to remind the Commission that although only a few states were members of the Commission — and most of those had, at that time, rather short-term interests in this industry — they were dealing with a resource in which all member states of the United Nations family had a legitimate interest, as did future generations of mankind. This was perhaps the beginning of FAO's acting as a kind of "guardian" on behalf of the world community in regard to the activities of international organizations which regulated regional and specialised fisheries. I found the whole problem most exciting, especially from the conservation point of view. We had a very great expert on our side in Dr. Sidney Holt.

Nutrition

I THINK it would be correct to say that nutrition is very much a Western science. The ancient systems of medicine in India or China did not separate nutrition as a subject to be dealt with by itself. It formed an integral part of the health of a person. Even in the West it is only in this century that the scientists have found it possible to reduce human food needs into the different elements — calories, proteins, vitamins, minerals — and to warn of the effects of deficiencies of any of these elements on the human system, especially in growing children.

The Hot Springs Conference (1945), which laid the foundation of FAO, gave much attention to this subject and this was reflected in FAO's Charter which emphasised this "raising of the levels of nutrition" as one of its prime objectives. The Hot Springs Conference proposed the establishment of National Nutrition Committees "composed of authorities in health, nutrition, economics and agriculture together with administrators and consumers' representatives, provided with adequate funds and facilities for the efficient conduct of their work, and having the authority to bring their recommendations to the attention of the public and to those agencies of government which deal with agriculture and the framing of economic and social policy". The essential point of this recommendation was that the National Committee or its equivalent should have the definite responsibility, and the means, to encourage and organize nutrition research, assess levels of nutrition, and speak with authority. It should be in a position to influence those responsible for national policies. Alternatively, personnel trained in nutrition should be added as full members to national Planning Organizations.

The reluctance on the part of governments in the developing countries to take such forthright steps appears to be due to several factors. First, the difficulty involved in changing the food habits of people, evolved over the centuries. I recall the tragic scene during the Bengal Famine (1942) when I saw a starving woman throw away the portion of wheat which was her share because it was not food to her! After that we organized mobile demonstrations, though it was too little and too late. Second, they feel that nutrition will follow from better production and improved distribution of available supplies; nutrition as such need not be given any special priority in their national plans. Third, the science of nutrition itself seems rather vague: what is described as good today may be condemned as a health hazard tomorrow. And now a fourth reason has appeared which is that the calorie intake must reach a certain level before the body can absorb proteins. In the developing countries, the calorie shortage is still the major cause of under-nourishment and malnutrition.

Coming now to FAO responsibilities, John Boyd-Orr, the first Director-General, laid down the ground rules: the study and appraisal of food consumption on world, regional and national bases in order to assess supplies needed for satisfactory nutrition; the study of physiological requirements for calories and nutrients and the practical application of the advancing knowledge; the study and application of specific measures to ensure effective use of available food supplies, with special reference to food technology, supplementary feeding and education in nutrition — these were the lines on which FAO's course was set. The first major contribution that the Nutrition Division made was when I commissioned the Third World Food Survey. For the final statistical calculations on the numbers of the under-nourished (and malnourished) in the developing world, the Statistics Division had to evolve new methods. This could only be done on the material so patiently worked out by the Nutrition Division on the preparation of national food balance sheets and food composition tables. Under the FFHC, work in nutrition was stepped up by the improvement of family welfare, including family nutrition; through Home Economics education and extension programmes; by direct assistance to governments in developing practical programmes in nutrition and home economics — these were some of the elements to receive special attention. Collaboration with other United Nations Agencies, in particular WHO and UNICEF, led to further development of new approaches.

The 1963 World Food Congress gave much importance to applied nutrition and strengthening national nutritional services, including technical institutes and training of personnel. Following the Congress, supplementary feeding programmes were extended in cooperation with UNICEF, and were also initiated under the newly created World Food Programme. Particular attention was given to the planning and development of school-feeding programmes and programmes for feeding industrial workers. Here I would like to mention my relationship with Maurice Pate, Director of UNICEF. We had first met when he came with ex-President Herbert Hoover to India in 1946. Even then his quiet dedication to the cause of children had left an impression on my mind. When I came to FAO, our personal friendship laid the basis for our work together. As our cooperation developed, we found it necessary to set up a Joint Policy Committee.

One special development during my time was the establishment of a Joint Food Standards Programme with WHO, its principal organ being the Codex Alimentarius Commission. The purpose of this Programme was to elaborate international food standards aimed at protecting the health of the consumers, ensuring fair practices in the food trade, harmonising definitions and requirements for food additives, hygiene, pesticides, contaminants and labelling. Its work also included coordination of all food standards undertaken by international, governmental and private organizations. By 1967, the work of this Commission formed the basis of harmonisation directives of economic groupings such as LAFTA (Latin American Free Trade Association), EEC and COMECON. Steps were also initiated to commence a programme of food control assistance to member countries which now form a major part of FAO's field programmes and projects in nutrition.

However, it must be recognised that while the international concern for raising nutrition levels remains as strong as ever, the limits are set by what the national governments themselves are able to do. And, for reasons I have stated, a long road lies ahead before nutrition takes its proper place in national development planning.

Land Reform and Rural Development

WHEN I WAS a young member of the Indian Civil Service I joined what was called a Settlement Camp which introduced new entrants to the Service to the complicated land tenure system in Bengal. The British had introduced a system which ensured regular payment of revenue by the Zamindars to whom the land had been parcelled out, leaving it to them to work out their own sub-contracting system. The result was a Zamindar at the top of a chain of middlemen while the actual tillers of the soil were grossly exploited and oppressed. The Settlement operations were intended to review the actual holdings and ownership of the small plots of land every few years which took into account the changes in ownership by inheritance or sale. The first task was to measure the plots of land and record on the map any changes that might have occurred. I remember spending days on end in heavily flooded lands trailing the measuring chains.

Before India's independence there had also been Land Commissions to bring about a better system of land-holding in Bengal. I was therefore well aware of the nature of the problems relating to land reform before I came to FAO. The general approach to land reform in most countries which had taken it up had been to give the land to the tillers of the soil and abolish the intermediate landlords who were paid some kind of compensation. This was what happened in India and other countries in South-east Asia. In the USSR and China the land was summarily expropriated. In Japan, under the Supreme Commander's rule, the landlords were eased out and paid compensation. Mere redistribution of land without provision of the necessary ancillaries (such as credit, guaranteed minimum prices, marketing facilities, etc.), would not really bring any radical change in ownership. Most governments were either unaware or reluctant to meet these needs.

In 1966 the World Land Reform Conference was held in Rome and was sponsored jointly by FAO and the United Nations with the participation of ILO.* This gave me the opportunity to make a contribution to land reform. I said at that time:

> Since 1951 when the first World Land Reform Conference was held, land reform has been approached in many different ways by different countries. An objective, comparative appraisal of the different approaches, and the mistakes, the failures, and the lessons to be learnt for the future, is now called for.
>
> We must first avoid confusion over the purpose of land reform. The arguments for social and economic equity, for income earning opportunity, and for creating a dynamic agriculture are interwoven in a pattern that obscure the separate threads. Each argument provides a basis for prescribing change in land rights, each can be separately delineated and sustained, but each does not necessarily lead to the same course of policy action. The pursuit of equity in the distribution of land wealth is a political and social goal that involves strict rights to ownership and tenure. These rights must include enforced limits to land holdings and detailed provisions for controlling the transfer of land between living persons as well as between generations. The legal frame for equity, however, does not necessarily provide for economic growth. This takes place only if there is cultivator response to incentives of stable ownership to tenure.
>
> Again, land as a source of self-employment, or family or tribal employment, will usually involve, in all but the most populated nations, a distribution pattern that chiefly specified a minimum area consistent with minimum livelihood. In the heavily populated nations, population and arable land might be so out of balance that it might be necessary in such cases to foster a pattern of farming that allows the joint social control of land, but at the same time, retaining in some way the incentive of personal ownership.
>
> While reforms in systems of land rights that have either equity or income opportunity as their goal are in themselves very desirable for ensuring agricultural growth, certain institutional arrangements are essential, such as extension of well-researched and adapted technology which can appreciably expand individual farm output; extension of an organized market system through which the farmer can sell his product and purchase his requirements; availability of production requisites that will enable him to exploit the new technology; and a system of price incentives to make the adoption of changed technology attractive.
>
> Therefore, institutions, with all their physical and human appurtenances — for education and research, for marketing, transport and communication, for maintaining price stability, and

*The International Labour Organization.

for the supply and distribution of production requisites and needs — must be the focus for any discussion of land policy for growth. This is the meaning of integrated land reform.

In all these discussions on land reform in the past, the quality of the farmer in the developing countries must not be underrated. If the farmer is conservative in outlook, it is largely because he has very little margin to play with. His interest and desire for a higher standard of living must be the basis for the economic and institutional infrastructure that must be built. Only if an objective assessment is made of the potentialities and limitations of land policy can the Conference focus on the proper balance to be found in the use of national resources for developing the institutions to support agricultural progress.

It is impossible to have any ready-made model. In every country and indeed, in some cases in different regions of the same country, the particular environments must be taken into account. We in the United Nations do not believe that we can shape land reform policies for governments, for honest decisions have to be made by the governments themselves in the light of their particular history and need. We can only assist them by appraising their current situations and indicating possible alternate policies and the consequences which might be expected to result from them. Another important factor is that as in any other field of investment, a continuous state of uncertainty about the future of land rights should be avoided, as it must dissuade people from putting forth their best in terms of capital and labour. This, however, does not mean that once the basic structure has been settled and fully given effect to, there may be no further need for evaluation and adjustment. Such factors as increases in population and changes in mobility, occupation, etc. will have to be kept under constant review.

Since I made these observations in 1966 some developments have taken place. We have come to realize a little more clearly that land reform by itself does not go to the heart of the problem of rural poverty. The Green Revolution which was heralded with so much promise has been found to benefit only a small section of the wealthier people. At the same time during these past years there has been a rising discontent among rural people who have experienced a feeling of deprivation and denial which is threatening to disturb the political atmosphere of their countries. The national governments have become increasingly aware of the real nature of the problem which has been highlighted so dramatically by the massive migrations of the rural poor to urban areas to seek food and employment. The whole process has been further compounded by the increasing pressure of population growth.

It will be relevant here to refer to certain new ideas, or rather new expressions that are current these days. One is that Basic Needs should have first priority in all developmental plans. But what are Basic Needs? Some in the International Community would look towards the poorest sections of the people, while many national governments would like to place the highest priority upon building-up the infrastructure necessary to accelerate and maintain economic growth. In either case, who is to decide? Who is to be the final judge if any suspicion of neo-colonialism is to be avoided?

Another question is what is to be given the higher priority in development — Equity in the distribution of the national product or Growth in production, assuming that this growth would percolate down to the poorer sections in the long run by the normal economic process. The controversy has at times assumed the shape of Equity versus Growth. Clearly, this is not a question of black and white. If Equity is emphasised to the exclusion of Growth, it may well mean a spread of poverty over the whole population where the GNP is low. If, on the other hand, Growth is emphasised without regard to Equity, the gap between the few rich and the many poor would tend to widen with incalculable social and political consequences. Thus, the problem is not Equity *versus* Growth, but Equity *alongside* Growth. Both should be pursued. But Growth should not wait for the modernisation of the social structure of the developing countries, which alone could ensure full and genuine Equity.

As for Target Groups (that is groups of poor families in selected rural communities to be taken up for special treatment so as to make an immediate impression on their standard of living), the World Bank has for the last few years been concentrating much of its resources on this approach. Our own experience in FAO through the various studies we have made — the Mediterranean Study, the Indicative World Plan, the studies for our SOFA* from time to time — is that our aim should be a total pull of the community as a whole, and in that process, one or other of the aspects may be given due emphasis according to prevailing circumstances in any particular area.

All these ideas and experiences came up for discussion at the World Conference on Agrarian Reform and Rural Development (WCARRD) called by FAO in July 1979. The highlight of the Conference was the address of President Nyerere of Tanzania. His was the authentic voice of the rural masses of the Third World. He gave a luminous exposition of the problems. The Declaration of Principles and Programme of Action by the Conference set out in a clear and forthright manner how nations with a political will could readjust their policies and priorities in their future strategies for development (*see* Appendix VIII, page 327).

The heart of the problem remains how to involve the rural people in the development process. The idea of people's participation was a key

*FAO's annual publication on The State of Food and Agriculture.

factor in the Freedom from Hunger Campaign (FFHC). The FAO Conference Resolution of 1959 makes that quite clear. As the Campaign progressed (see Chapters 13 and 14), the part played by the people at the grassroots level proved a major positive factor. Perhaps our FFHC contributed to a movement which is now gathering strength.

Desert Locusts and other Pests

IN THE UNITED NATIONS system, coordination of measures to control pests has been the responsibility of FAO.

Desert locusts are a very serious problem and FAO first became involved in anti-locust activities as far back as 1951. The Desert Locust causes extensive damage to crops and other vegetation over an area of some 20 million square miles in Africa and Asia. FAO has established Regional Commissions in north-west Africa, the Near East and south-west Asia, to support national operations. Until the 1950s, control measures could not effectively combat the enormous swarms. But the development of specialised techniques using aircraft, concentrated insecticides and improved strategies, now provide the means to prevent the recurrence of major plagues. There has been no lack of interest on the part of the scientists and I saw them at work in several research centres I visited.

By instituting a strict watch over the main breeding grounds — Sudan, Ethiopia and Somalia in Africa; the Kingdom of Saudi Arabia and the two Yemens in the Middle East; and India and Pakistan — it was possible to maintain effective control for nearly sixteen years. In 1978 this control broke down because the breeding grounds in Ethiopia and Somalia could not be reached because of the state of war between those countries. The locusts threatened the countries of West Africa and north-west Africa, apart from their normal sweep eastwards which takes them as far east as India.

I had attended meetings in FAO before I became Director-General which led to the institution of the Desert Locust Control Committee. I have to place on record that this problem has always received very close attention from the Western countries who helped with finance, machinery and other requisites in the fight to eradicate this menace.

As for animals, I recall that in working out the scope of the World Food Programme in 1962 I had specially included epizootics among livestock and poultry as "emergencies" requiring as much attention as emergencies directly involving human beings. I understand that some hesitation has developed during the past few years among members of the Governing Body of the World Food Programme over this matter. I hope that considering the kind of relationship that exists between man and his animals in some countries, the original position will be maintained.

The World Food Conference in 1974 adopted a resolution calling upon FAO to intensity its work on trypanosomiasis — sleeping sickness and other diseases, caused by a parasite in the blood of animals and also affecting humans — which is a problem of great magnitude over millions of square miles of Africa. The World Food Conference specifically noted that trypanosomiasis control should be considered as the first step in an integrated plan of economic development. This first step should be followed by projects covering appropriate land, water, and forestry conservation and utilisation including pasture improvement, livestock management, animal health, livestock marketing and processing, as well as training in the various fields. FAO was to be the UN Lead Agency not only for coordinating ongoing operations involving international organizations, specialised research institutes and various bilateral and multilateral supporting agencies, but also to take responsibility for a sustained long-term programme of integrated economic development. No doubt, it will take generations to complete a programme of such magnitude, or even to bring the problem under a reasonable measure of control. But the stakes are high. Any real progress would be of incalculable benefit to the future of the great emerging continent of Africa.

Emergency Relief

IN THE PAST emergency relief for disasters arising out of natural causes was the sole responsibility of the national governments, or of the Colonial powers for their dependent territories. An attempt was made in India over a century ago to deal with such disasters in a systematic way. The Indian Famine Code remains more than an historic relic. The concept of "Food for Work" now used in many countries originally took shape under the Indian Famine Code.

The emergence of the League of Nations first provided the opportunity for common international action. But when discussions were held on problems of food and nutrition (in which one of FAO's pioneers, Viscount Bruce, took an active part), organized international action for emergency relief found no champion. It was only after the United Nations came into existence that the international community began to take an active interest. In the early fifties, by formal resolutions, both UN and FAO recognized the problem and their responsibility. FAO was to be the UN Agency for organizing international assistance in emergencies affecting not only human beings but also livestock on which human beings depended. The Hungarian relief operations in 1957, and the international appeal for funds and food supplies to meet the Bihar famine in India in 1964/65, were undertaken by me as FAO's Director-General under these mandates.

This was the situation until the end of my term of office in 1967. Since then, with many new nations gaining independence and coming forward as members of the United Nations, the need for emergency assistance on a broader basis to include earthquakes, flash floods, tornadoes and crop failure due to drought has been given prominence. The United Nations has set up a Centre (UNDRO)* to deal with emergencies in a more comprehensive and systematic way. The new approach involves action in diverse fields such as warehousing at ports of entry, provision of jeeps and trucks to facilitate internal distribution, organized shipping, and other measures.

The expansion of UN responsibility for emergency relief has brought some interesting new initiatives. The role of the World Food Programme (WFP) as a coordinator of all food aid, whether multilateral or bilateral, has been strengthened: "The Programme shall, within the framework of emergency assistance cooperation in the United Nations system and in accordance with appropriate recommendations of the United Nations and FAO, seek to ensure coordination of emergency food assistance." The governing body of the Programme, however, is divided on the share of the Programme's resources that should be devoted to emergency purposes — emergency relief often has a strong political element as far as the donors are concerned who usually are the developed countries.

The setting up of the International Emergency Food Reserve (IEFR) in 1975, with the target of 500,000 tons, replenishable annually, has strengthened FAO and WFP in this work, though the target has yet to be reached.

Another development took place when the Secretary-General of the United Nations entrusted overall responsibility to one UN agency for coordinating all emergency assistance including what might formally fall outside the scope of that agency. The first major undertaking for FAO under this arrangement was the Sahelian relief operation in North Africa. This was justified at the time by the peripheral nature of non-food aid. The experiment proved successful, particularly in integrating multilateral and bilateral aid.

Unfortunately, this arrangement does not appear to have worked too well in all cases. The relief operation in Kampuchea has presented a picture of hesitation, inaction and misinformation. UNICEF has been widely blamed. Much of the confusion is no doubt due to the complex political conditions in and around the country and the years of disorder due to external and internal aggression. One relieving feature in this tragedy appears to have been the initiative to supply fresh seeds, fertilizers and other agricultural production requisites to one part of the

*United Nations Disaster Relief Office.

country so that at least some of the afflicted people could look forward to the next harvest.

As one whose lot has been to deal with many disasters, both natural and man-made, I might say that while the present UN machinery for emergency relief appears generally sound, its working seems to call for a fresh review.

I often wonder if there is sufficient delegation of authority to enable urgent policy decisions to be taken at the operational level. Should there be more flexibility in the choice and role of the lead agency, especially when the situation is prolonged and changing in character? Is sufficient importance given to maintaining a close operational relationship between the UN organizations on the one hand, and the non-governmental organizations, national and international, on the other?

In the Kampuchean case, these organizations appeared to be running on parallel lines and, indeed, sometimes in conflict with each other. One of the lessons we learned from the Freedom from Hunger Campaign was the value of close cooperation between the interested and involved parties.

I am now a more distant observer of the human tragedy that has unfolded in several parts of the world. Large-scale movements of population in search of food and shelter is one problem. There is now the further problem of a large-scale movement of people in search of a better way of life. In some areas, the two streams merge into one. What is most disturbing is that this agonising human tragedy does not dissuade the Super Powers and their allies from coming in to play their own self-interested geo-political games!

Taking the case of the Vietnam Boat People, it was a generous and caring gesture to rescue them from the seas and find them temporary accommodation in relief camps. But no attempt was made at any stage to extend economic and other assistance to these disturbed people to encourage them to stay and find a more assured way of life in their own land. If politics prevented bilateral assistance, could the multilateral agencies and non-governmental organizations not have been entrusted with that responsibility?

Chapter 23

UN and FAO: Some Recollections

Secretaries-General I have known

THE THREE Secretaries-General I came in personal contact with were Trygve Lie, Dag Hammarskjöld and U Thant. I met Trygve Lie when I was in the Indian Embassy in Washington and I came in contact with Dag Hammarskjöld and U Thant during my days in FAO.

It is difficult to say what makes an ideal Secretary-General. It is one of the key positions in world affairs and yet one of the most difficult. The UN is charged with responsibility for both War and Peace, "to save succeeding generations from the scourge of war" as well as "to promote social progress and a better standard of life in larger freedom", and the Secretary-General is the "chief administrative officer" whose responsibility it is to pursue those noble goals. To have in one person all the qualities and qualifications to discharge such wide responsibilities, and in addition "keep under review matters which in his opinion may threaten the maintenance of international peace and security", is asking a very great deal indeed.

The Secretary-General is usually one of three kinds — a super-administrator keeping the machinery well-greased and smoothly functioning and at the same time providing at least the minimum leadership to get the member countries to discharge their responsibilities; the second kind, a super-political scientist who is especially active in carrying out the objectives of UN in the maintenance of international peace and security, giving only secondary attention to other matters; and the third kind, a super-economist who gives his personal attention and leadership to social and economic problems which in fact lie at the root of the world order. Trygve Lie was probably one of the first kind; he kept the UN in its formative stages running smoothly without taking too active a part in economic or social matters. Unfortunately, he fell foul of one of the super-powers because he took an active interest in the question of a UN peace-keeping force for Korea. He was a warm-hearted, outgoing personality, and in retrospect will probably be regarded as one of the best and most successful Secretaries-General that the UN has had. I met him again in 1958 in Oslo; the years had not dimmed his robust optimism nor his humane outlook on world affairs.

Dag Hammarskjöld, a somewhat remote and self-centred man, cold with his colleagues, was a trained economist (I remember Gunnar Myrdal once telling me that he was one of the examiners of his Doctorate thesis at Uppsala University) and would probably be classed in the second category. He was not a politician but was intensely interested in political matters. He had a sense of mission and wanted to make the World Body really effective. He wanted to transform the position of Secretary-General into that of Chief Architect of the UN. It was this sense of mission and his surpassing dedication which he sometimes allowed unconsciously to exceed his constitutional powers that finally sealed his fate. His last days before his death in a plane crash going to the Congo bear testimony to both his qualities and his limitations.

My acquaintance with Dag Hammarskjöld began with a serious misunderstanding. This was over the matter of relief to Hungary after the Soviet invasion. I got an SOS from the Hungarian Government that their potato crop had been totally destroyed and unless they got fresh potato seeds immediately the food situation in the country would soon be serious. Under resolutions of both the General Assembly and FAO, the Director-General of FAO was responsible for emergency relief on behalf of the UN system. I decided to send Dr. Wahlen, head of the Agriculture Division, to Hungary to assess the needs. (Dr. Wahlen was an eminent man in his own field. During wartime, it was the "Wahlen Plan" that saved Switzerland from what could have been a serious food situation. He became my Deputy and later on in consultation with me and with my encouragement left to stand for election in his country, ending up as a Federal Councillor and then in his turn the President of Switzerland.) As soon as Dag Hammarskjöld heard of my plans he sent me an urgent message not to proceed and await his decision. I did not see that I had to conform since I had statutory responsibilities independent of him in the matter. I asked Dr. Wahlen to continue on his mission and he did. He crossed the border with some difficulty but finally was able to get the information we needed to mount a relief campaign. I approached the Netherlands and Germany among others in western Europe who were likely to have the varieties of seeds that would suit the conditions in Hungary. Only when the special trains loaded with potato seeds started rolling across Europe did Dag Hammarskjöld come round and instruct his representative in Geneva to cooperate with Dr. Wahlen. There were some other incidents over this operation but I need not go into them here. It was only much later that I learnt the reason for Dag Hammarskjöld's attitude — he had been refused permission to visit Hungary immediately after the invasion. However, though this beginning somewhat chilled our relations, it did not dim my admiration for his great pioneering spirit which he showed all through his term of office.

Dag Hammarskjöld had one hobby in common with me — we were both fond of taking photographs during our travels. Once he had been to Nepal and had taken a series of photographs from a plane flying across the Himalayas. I had followed him a few weeks later and had done the same. When I met him in the UN, he showed me his photographs which were almost professional in quality. Soon after his death, one of these photographs was published by the London Times, filling the back page. The high rugged range with the peaks in eternal snow somehow portrayed his image!

U Thant, however, did not fall into any of these categories. He was a man apart. Above everything else, he had some of the rare qualities of a true Buddhist — gentle, contemplative, generous and at the same time detached in spirit. He was not an administrator, not an economist, not a politician. But his detachment in that position was a rare quality. He once told me that he was very fond of watching boxing on TV (an unusual thing in a Buddhist!) but never favoured one side or the other. I thought this brought out his true nature. Unlike Dag Hammarskjöld he made a point of remaining one step behind the event. His reactions to critical situations were slightly delayed, no doubt by choice at times, but his temperament had something to do with it. In the only instance where he acted swiftly, withdrawing the UN peace-keeping force in the Near East, he was criticized for acting too quickly since he did not have time to clear his position with the Security Council. The consequences of his decision were disastrous leading to the Arab-Israeli war of 1967. His part in defusing the Cuban crisis has been handsomely acknowledged by the then British Prime Minister, Harold Macmillan, in his Memoirs.

U Thant was a close friend of mine. At the Fourteenth Session of the FAO Conference (November 22, 1967), one of the valedictory addresses was the one delivered on his behalf which had these words:

> One of his most notable achievements has been to concentrate attention on the world problem of hunger and to spearhead new action including the institution of the Freedom from Hunger Campaign to overcome it. He has succeeded because he has proved himself to be not only a man of deep moral purpose but to a remarkable degree a man of action. In all that he has done, moreover, Dr. Sen has understood that the international community is not only merely composed of governments, but also, to use the

> terms of the UN Charter, of 'the people of the United Nations'... He has realized that this means a bond between the organized international community and ordinary people everywhere, and thus, in large part due to his dynamic and restless energy, who does not now recognise the dangers that mass starvation may hold for the world and who does not now recognise the possibility for international action to cope with this problem?... For me, it has been a privilege to know and work with Dr. Sen, both directly and through the Administrative Committee on Coordination (ACC) during all the years I have been Secretary-General. Not only the United Nations but the whole of the United Nations system owes much to him, and we count heavily on him in the years to come to benefit from his counsel and his support.

I cherish this tribute above all others and if I quote his words which paid me homage, it is more to illustrate his ideas and his great and generous spirit.

We know that he found time every day to practice contemplation. This for him was not an empty ritual but an expression of the true Buddhist spirit which encourages a person to rest the mind for a while on eternal things and is thought to be a spiritual support for withstanding the pressures of the daily material life.

The UN Family Network

NONE OF THE Secretaries-General I had worked with was really interested in administration, i.e. in their secretariat or its structure. The secretariat was left more or less to itself. There was never any serious attempt to rationalise the structure of the organization. Departments, Divisions, Offices, Bureaux, Units, were set up as and when required. Once established they remained. The result was a good deal of confusion. The lines of command were tangled and the relations between one senior officer and another were more personal than organizational. Another result of the preoccupation of Secretaries-General with political problems was that some of the essential responsibilities of the Secretary-General were taken over by governmental committees set up by the General Assembly which in FAO had always remained with the Director-General.

The area of most difficulty in relations between the UN and FAO is the UN's attempt from time to time to define its relationship with Specialised Agencies including FAO. The UN Charter does not envisage any direct interference with the working of the Specialised Agencies: the Economic and Social Council is to exercise its overall coordinating function "by consultation with and recommendation to" the Agencies. Yet, ECOSOC and the General Assembly from time to time have embarked on exercises contrary to the spirit of these provisions of the Charter. The first spurt of activity came in 1948 when under the dynamic and somewhat masterful leadership of Sir Ramaswamy Mudaliar (India) as Chairman of ECOSOC, agreements were 'imposed' on Specialised Agencies which ran quite contrary to both the spirit and the letter of their charters. It soon, however, became clear that the UN with its disorganized secretariat and scramble among senior officers for greater jurisdiction had undertaken too much and the agreements were not acted upon in practice.

Coordination has always been an important fact of life for the UN system. At every ECOSOC debate, coordination has been one of the principal items on the agenda, and every Specialised Agency spokesman has had to expound how the Agency was cooperating with the rest of the UN system. The intention behind this theme, of course, was that duplication and overlap should be avoided. We now seem to have come to a stage where coordination has assumed a new meaning. Every General Assembly resolution these days insists on every UN Organization producing studies, attending meetings, taking operational action to implement so-called global strategies. The result is that all UN Organizations are forced to do work and spend resources on work which sometimes is of low priority to one or the other. Nowadays, it is not the avoidance of duplication but centralisation which seems to be the propulsive force. What is being lost sight of is that this kind of centralisation can be destructive of the initiative and freedom of action which gives life to the Specialised Agencies.

This is an unequal fight. The Member Governments are represented in the UN by their Foreign Offices, and in the national government the Foreign Office is always politically more powerful than the Department or Ministry of Agriculture. The Member Governments sometimes fail to realize that there must be some meaning in the concept of Specialised Agencies apart from the UN itself. In their own country, they have separate Departments or Ministries for Agriculture, Health, Education, Labour, because these subjects need specialised attention. In the case of the UN Specialised Agencies, each has a separate Charter written by the Governments themselves and have independent governing bodies and independent Executive Heads and budgets. Disturbing this arrangement can only harm the interests of the Member Countries themselves. They should exercise a little more circumspection in this matter and not destroy the initiative and freedom of action of the Specialised Agencies. Many new ideas on development come from these agencies and FAO has been in the forefront. Coordination cannot in itself be the ultimate objective; in the quest for coordination, the innovative spirit of the Specialised Agencies and the spirit to explore new ideas for action should not be sacrificed.

Some New International Institutions

THE 1974 GENERAL ASSEMBLY resolution calling for a World Food Conference came in the wake of the oil embargo on the USA and on other industrialised countries. The resolution was moved by Dr. Henry Kissinger, Secretary of State of the USA. It would not perhaps be far from the truth to say that the real motive behind this resolution was to bring to the notice of all that while some countries had a monopoly of oil the US had food to bargain with. Apart from any other criticism this move was a serious breach of the long-established tradition that food should be kept out of politics. The sad thing is that it did not occur to any of the developing countries, and in particular to FAO, to bring this aspect of the matter before the Assembly. The 1961 General Assembly resolution for provision of food surpluses to food deficient people through the UN system would have appeared in quite a different shape had I not taken the steps that I did to place the whole matter in a proper perspective.* Even if FAO was unable to forestall the political war in the UN in the present case it does not appear that the FAO Conference in the following session took any steps to advance its position in the matter, though the FAO Secretariat must have taken a considerable part in drafting the technical resolutions, almost all of which related directly to FAO responsibilities and could have been passed by a FAO Conference.

The lack of FAO leadership found expression in particular in the ready acceptance of the new institutional arrangements that by-passed FAO almost completely. A World Food Council was to be created consisting of members nominated by ECOSOC and elected by the General Assembly to carry out "an integrated multi-disciplinary approach within the framework of economic and social developments as a whole". It was to be the political organ of the UN system on the argument that the problems involved could not be entrusted to FAO alone. If one goes back to the early days of FAO, one would find that the FAO Council was supposed to do more or less what is now contemplated for the World Food Council though the wording of the WFC resolution is somewhat more inflated than we in those days were accustomed to. In fact, one of the recommendations of the high-powered Preparatory Commission set up to examine Boyd-Orr's proposals on a World Food Board in 1946 was a Council of FAO to act between the sessions of the Conference, calling it a "World Food Council". Rather than an entirely new body floating in mid-heaven, could not the restoration of the original idea of making the Council membership of high policy level and, in addition, having one session in the year specially subject-oriented, have served the same purpose and at the same time given it the added strength of the whole of FAO behind it?

*See *History of the World Food Programme*, in Chapter 20, page 197.

A second innovation was the creation of an International Fund for Agricultural Development (IFAD) as a Specialised Agency outside FAO. I have already commented upon IFAD and its terms of reference as compared to those of my original proposal for the creation of such a Fund (FFRP).*

Another innovation recommended by WFC was to re-constitute the Intergovernmental Committee of WFP "to help evolve and coordinate short-term and long-term food aid policies in addition to discharging its present functions". This again could have the effect of diluting the overall control that FAO exercises under the original Intergovernmental Committee arrangement, unless there is full understanding between FAO and this new body all along the line.

The CGFPI (Consultative Group on Food Production and Investment in developing countries) was a misbegotten offspring of the WFC and, happily, breathed its last after various attempts at artificial respiration.

The one general comment I would make is that the member governments should be warned against the growing tendency in the General Assembly and ECOSOC to pass resolutions setting up new institutions to do the work that could well be done by the existing agencies. If the work was not being done by these agencies it was because the member governments themselves were not giving them sufficient attention to make them work. As for the industrialised countries, they should take heed of this development if for no other reason than the one closest to their hearts — to keep down the ever-increasing budgets of the UN system.

However, when all is said and done, food is now in the political arena and nothing will redress this very unfortunate development.

The FAO Conference: Some Personalities

THE FAO CONFERENCE meeting once every two years brings together national representatives from all over the world — from the most populous states to remote islands of a few hundred thousand people. Every delegate comes as an equal. In the sixties with the break-up of the old Empires, there was a large accession of independent states. The most striking were the Africans who came in the consciousness of their new identity and were the most colourful in their different costumes. On one occasion I remarked to a tall handsome Nigerian who had his wife beside him how it was that he had such resplendent headgear and neckpiece when his wife did not have anything to match them. He said, "Look at

*See The World Bank, in Chapter 21, page 214.

the animal kingdom, the males always have the pride of manes, tufts, plumes and tails!"

It is also interesting to observe the voice and the delivery of delegates in different languages. You have the panorama of the whole world under one canopy.

Every Conference is notable for one reason or another — in the subject-matters discussed, in the clash of group interests, in the examination of new ideas or initiatives brought up either by the delegates or by the Director-General. Except on rare occasions the atmosphere is usually marked with a rare courtesy and consideration. But the Conference which to my mind was the most unforgettable was the one in 1959. It brought home to me the feeling of the flowering of the post-war idealism which had brought FAO into being. A record of my thoughts at that time had these words:

> The year that had just ended marked the close of a decade. Whatever else might be the verdict of future historians on the 1950s, no one would dispute that this period saw the coming of age of the United Nations. Nor would it be denied that the fifteen years since the end of World War II, in striking contrast with the same period following World War I, had seen an unparalleled record of reconstruction and economic recovery in the war-ravaged countries — a development in which the United Nations and its technical agencies played no mean part. As Mr. A.J.P. Taylor, the Oxford historian, had so aptly put it: "After the First World War, everyone believed in the League of Nations and it did not work. After the second war, no one believed in the United Nations and it is working very well. We are entering Utopia backwards, constantly surprised that the future turns out so much better than we expected."
>
> FAO could not fail to draw fresh vigour and strength from this steady advance made by the international community towards maturity. The fifties witnessed lively adventures in sovereignty and experiments in democracy in many countries which had just won their political freedom. One outstanding lesson of all these strivings was the realization that economic progress and social security took longer and were more difficult to achieve than the paraphernalia of a sovereign state. The energies of the economically less-developed countries had, therefore, been turned increasingly towards an intensified search for economic prosperity. Planning for development had been the watchword of national effort. Whatever else happened, the sixties were going to witness a progressive and intensified mobilisation of national and international resources, and of modern technology and organization for combatting poverty and want wherever they existed. The long march towards the attainable Utopia had begun, and there was no going back. On this route one could only go forward.

> This spirit was abundantly manifest in the deliberations of the Tenth Session of this 1959 FAO Conference. Without doubt the Conference was a great success, whether one looked at it from the viewpoint of the spirit of harmony and understanding that prevailed throughout the entire Session or from that of the smooth and efficient manner in which the delegations accomplished their onerous tasks. Indeed, this Conference, in the opinion of many veteran delegates, was perhaps the best-conducted in FAO's history. The credit for this success must go to the delegations who so ably represented our Member Nations. The Secretariat also felt rewarded for the efforts made in planning and preparing the Conference. Nevertheless, the most significant feature of this Conference, to my mind, was the mood of courageous leadership which pervaded its deliberations.

My impressions of some of the Chairmen of the Conference and the Council I had worked with were varied. Since the Conference met only every two years and the function of the Chairman was to preside over one session only, there was little scope for any long-term acquaintance between them and the Director-General. So long as FAO was located at Washington, the Agriculture Secretary of the United States (which was the host country), presided. Of these Chairmen, I came to know only Charles Brannan. I have already described my encounter with him over the transfer of the FAO Seat to Rome.* It was the usual practice in Rome for the Director-General to take the initiative to suggest a name in advance usually by rotation of the main regions, and submit it through established procedure to the Conference for formal adoption.

In 1961, because of the emergence of new independent nations in Africa, my choice fell on Julius E.Cooper, who had been attending the Council meetings as the Delegate of Liberia. After I had made my choice known to the Liberian Government, I found them none too happy as Mr. Cooper was not of ministerial rank. (They were very protocol-conscious in Liberia in those days.) For the 1963 Conference I thought an Asian should have the chair and my preference was for Ambassador Nasu of Japan whom I had known as a Council Delegate. I discovered too late that he had a rather exaggerated notion of his own importance as Chairman of the Conference. When the tragic news of the assassination of President Kennedy came through, the Conference delegates expressed their shock and grief. When I told the Chairman that I would like to speak, I found him unwilling to give me the floor because he thought that the Conference was of Member Governments and I was only a functionary! However, I brushed him aside and spoke to the assembled delegates. After that, Nasu and I never quite recovered our old relationship.

*See Chapter 10, page 128.

The 1965 Conference was chaired by Mr. Maurice Sauvé — Agriculture Minister of Canada whose warm friendship I still enjoy. My last Conference was presided over by an old friend of mine whom I had known as a colleague when he was Ambassador to the US, Mr. Joseph Winiewicz of Poland, one of the most experienced and liberal-minded men I had come across in my career as a Diplomat. I am happy that it was his signature which appears in the parchment of Citation of the Vote of Thanks Resolution passed by the Conference in 1967.* It hangs on my wall reminding me of those exciting days!

The FAO Council: More Personalities

WHILE THE COUNCIL does not have any of the colour and excitement of a FAO Conference, it serves as an essential prelude. With the help of its ancillary committees, specially those of Programme and Finance, it lays the ground work for the policy and programme decisions of the Conference. However it has to be admitted, with some regret, that as the years have gone by, the high representation envisaged for the Council in its original concept as "The World Food Council" has not always been maintained. The great figures like Professor André Meyer and Viscount Bruce are not easy to replace. The President of the present World Food Council speaking to the 1977 FAO Conference took pride in his Council having been attended by thirty-one mythical figures going by the title of Plenipotentiaries and the Conference itself had no less than ninety-one Plenipotentiaries. There is no magic in numbers, it is the quality which counts.

To recall some of the personalities who gave life to the Council during my office — of the Chairmen of the Council, the two I admired most for their intellectual qualities were Viscount Bruce and Professor André Meyer. André Meyer was one of those truly learned professors greatly respected by his students — gentle, kind, understanding — a "teacher" in the best traditions. I was happy to have been given the opportunity to honour his memory with the institution of the *André Meyer Scholarships* for deserving young research students. The second Chairman I came across was Viscount Bruce. I hope I will not be misunderstood when I say that he was a class apart, an international figure of unparalleled experience. I remember him sitting through a Council session and at the end of a three-hour debate summing up the salient points of the discussions without any notes and not missing anything. At a party at my residence in Rome towards the end of 1952, he told me he would be leaving the FAO Council soon and invited me to take his place. I said

*See Appendix IX, page 331.

that I would convey his approach to my Government. My Government, however, decided to put up the name of Sir Ramaswamy Mudaliar, a powerful figure in the UN, having been the first Chairman of ECOSOC. But not being known in FAO circles, he lost to Josue De Castro (Brazil). I found De Castro an agreeable man. He came with the reputation of being a 'leftist' — in those days an unpardonable sin — and had to struggle all through his term to maintain his position as Chairman. He was followed by Hasnie (Pakistan) who made a fine Chairman, very fair, full of good humour and camaraderie. He had the reputation of being a very liberal-minded man which was a difficult attitude to maintain in those days of upheaval, following the partition of India. When Chairman of the Council he was also Governor of the State Bank of Pakistan. When I visited him at home I was shown a fresco on the wall depicting great men of the East. One of them was Tagore, the Poet of Bengal. Adulation for Tagore was quite a heresy for a Pakistani at that time.

I always stayed with him whenever I passed through Karachi and enjoyed his unlimited hospitality. George Haraoni (Lebanon) succeeded him but before he could join us was overtaken by a fatal illness. Just before his illness we had spent some happy days together among the Cedars of Lebanon in the depths of winter. Our mountain hut was buried in a heavy snowdrift but inside there was warmth, light and laughter. Those all too few unforgettable days! Edouard Saouma, the present Director-General was in our party and I recognized then that he had a promising future.

Among the delegates there were some who for one reason or another left an impression on my mind. First, my friend, the Maharajah of Patiala, who headed the Indian Delegation in 1963 and whom I specially requested from the Indian Government in 1967. A very handsome man whom I had known since his younger days, he was over six feet four inches, of straight bearing, a well-kempt beard and a turban which was an ever-changing work of art. He was capable of dazzling the secretaries at the Conference who could not take their eyes off him and proceedings seemed to slow down in that department when he was around. For sheer intelligence I found very few to equal Edgard Pisani, Agriculture Minister of France. I remember him and Soames of the United Kingdom being locked in a debate at the 1963 Conference. Edgar Pisani was the winner but Soames was a close rival. Soon after, Edgard Pisani left his government and joined commerce. I thought his departure was a great pity.

Freeman, the US Secretary of Agriculture, was very fond of the expression 'nuts and bolts'. At the World Food Conference in 1963, over which he presided, I remember I counted the times he used that expression. Despite our differences, he remained a friend and gave me his fullest support at a time of need.

El Ghorfy and Sbihi of Morocco were my staunch supporters all through. I had met King Mohammad V on my first visit to Morocco. His son and heir, King Hassan, extended his warm friendship — I visited his country several times to oversee some schemes we had started — and at my request donated the expenses to refit the main Dining Hall in FAO, named after his father.

I have already mentioned Dr. Mansholt of the Netherlands and Dr. Lübke of Germany who, in different ways, were always a source of strength to FAO.

Women on the International Stage

IN MY TIME AT FAO, I came across several women leaders I shall always remember. I have already mentioned Alva Myrdal. In FAO, Dorothy Jacobson (Asst. Secretary, Agriculture Department, USA), proved to be my 'right-hand person' through many of the tortuous debates in Council and elsewhere. Princess Souvanna Phouma (Laos), one of the finest products of French culture who spoke French better than most Parisiennes, had her feet firmly on the ground. I remember my visit to Vientiane when she drove me round the nearby villages to show the work being done under the FFHC. She died a few years ago and it was a personal loss. Madame Lopez Mateos, wife of the Mexican President, was another outstanding woman whose social work was an example to other countries. Barbara Castle, Minister of Overseas Development in the UK, came at the head of her delegation in 1965 and we became friends. Her presence was the highlight of that Conference. Finally, to close this short list, Madame Marie-Therèse Basse (Senegal) proved a most devoted worker for the FFHC in her country and whose services FAO has since recognized by the award to her of the much coveted *CERES* medal.

We hear a lot these days about women and their place in society, government, and international organizations. In national governments, there have been outstanding women in the past as well as in recent history. In the international organizations, however, men still guard the citadels.

Why should a woman not be Secretary-General of the United Nations? Women of the world, surely you could find among your ranks a Margaret Thatcher, an Indira Gandhi, a Barbara Castle, an Imelda Marcos, a Simone Veil? I feel sure that once such a formidable candidate appeared on the scene, none of the super-powers would dare oppose her. Perhaps we can look forward to a woman rescuing us from the sorry state which men have brought on the world.

One World?

I HAVE ONE observation to make before the curtain is finally rung down on the scene. Why are the developed countries so allergic to budget increases of FAO even when the case made is so compelling? The reason cannot be financial since the developing countries also share the burden on an equitable basis, on an agreed formula worked out by the United Nations. The developing countries do not complain. True, the developing countries derive ostensible benefit from FAO programmes. But the developed countries also benefit in a different way from FAO's long-term studies in that they can adjust their production and trade policies. In any case, after exploiting them for so long, is there not some moral obligation now to help reduce the income gap between the developed and the developing countries.

It seems to me that the reason for this negative attitude is more psychological and political than financial. Politically the developed countries are now in a minority in all the UN organizations. Any resolution requiring a two-thirds majority can be passed by the developing countries without much difficulty. The developed countries are, therefore, on the defensive and are reluctant to give more money or more power to FAO or any other Specialised Agency of the UN Organization. The present insistence on channelling all developmental aid by voluntary pledge through the UNDP could well be a way of having a greater say in spite of the voting weakness. The *Helms Amendment* passed by the US Congress last year, making US contributions to the Agencies conditional on none of it being used for technical assistance, was primarily aimed at FAO's new initiative — the Technical Cooperation Programme (TCP). This is an expression of the same uneasy and insecure state of mind. I also believe there is a fear that once the principle of technical assistance through an "assessed" budget is allowed to come in the course of time this may take the form of an "international tax" which with the loss of their voting majority may lead the developed countries to providing a "bottomless pit".

Here there is clearly a crisis of confidence which needs to be faced. It has always been my feeling that consensus and not confrontation should be the basis for discussions in the UN system. Unless an atmosphere of mutual confidence and reliance is restored, sooner or later the United Nations could go the way of the League of Nations. There must be a spirit of give and take. The developed countries must understand that they have much more to give than the developing countries to redress the present economic imbalance.

Social Life

IT WOULD BE incomplete if I filled these pages only with my FAO work. I came to Rome in 1950 and, with a break of only one year, stayed on until 1954 as India's Ambassador. In that capacity it was a joy for me, my wife and my three growing daughters to see Italy from Venice to Sicily. We saw practically all the works of art — but this is too extravagant a claim, for who can ever exhaust the treasures of Italy! When I took over at FAO I had little time to continue these adventures but my wife who is a keen student of art in all its forms kept up the interest. The famous Book Club for the wives of Officers was started in our time. At every meeting one member would read an essay on a selected subject on the art or architecture of Europe which was followed by discussions. My wife used to tell me of the high standard of some of these essays. Sometimes excursions were arranged for the study of special places.

Official cocktail parties were the bane of our lives. During conference time, every country used to give a gala party to establish its identity and importance and I could not miss any of them without creating a diplomatic incident. On one occasion I had to attend five parties between 6 and 9 p.m. in different parts of a city which was choked with traffic. I decided that I could satisfy my hosts if as I entered I took a drink in my hand and walked briskly through the reception hall and disappeared at the other end. I managed to make my appearance at all of the five parties but arrived late for the last one which was a dinner party in Via Appia Antica on the other side of town. After this experience I persuaded the countries to accept my proposal of having these parties by regions and as far as possible in the FAO Restaurant.

We had a very modest home for the Director-General of FAO but my guests did not seem to mind the humble abode and thanks to the warm welcome and generous hospitality of the hostess, the guests always seemed to be well pleased. Lately, one of the delegates referred to these parties — some twenty years later — at the ceremony of Award of the Agricola Medal to me in FAO. In summer time we used to entertain our guests on a small terrace which had a tiny patch of grass in front. On one occasion we had invited Cardinal Frings of Germany* to dinner when my butler, quite uninstructed by us, went out with lighted candles to escort the Cardinal in from his car!

*Cardinal Joseph Frings proved a powerful force in promoting the Freedom from Hunger Campaign by associating the Catholic Church in West Germany and Catholic organizations working in different countries all over the world with the Campaign.

Chapter 24

My Travels

India's Influence Abroad

SAID BEN SIRA in 190 BC, "A man that has travelled knows many things." Whether I know many things or not I have travelled widely. I have already touched on some of my travels during my diplomatic career in the service of my government.* But my greatest opportunities for travel came when I assumed charge of FAO in 1956. During the next eleven years I went to practically every country in the world.

During my travels I first visited the capitals to discuss the problems in hand with the authorities, then I liked to go out into the countryside to study the life of the people and see objects of historical and cultural interest. In these travels I was always fortunate to have the fullest cooperation of the authorities who willingly provided all the necessary facilities.

I do not propose to deal here with the food and agriculture problems of the countries I visited but very briefly to mention a few of the "many things" that I learnt in the course of my travels. The regions which interested me most were South-east Asia and the Middle East. Naturally, I looked at these countries from the perspective of an Indian.

How wrong was H.G. Wells when in his monumental *Outline of History* he wrote: "One might write a history of India coming down to 400 years ago and hardly mention the sea." Yet, between the first and the eleventh centuries, there was active trade and travel between

*See *On the Road*, Chapter 7.

India and the south-eastern countries. This led to the spread of Indian religion, Indian culture, Indian laws and Indian forms of government over wide areas. It was against this background of history that I saw some of the ancient monuments and ruins of South-east Asia.

The Buddhist stupa of Borobudur in Java was raised by the Sailendra dynasty whose empire comprised the whole of the Malaya Peninsula as well as Sumatra, Java, Bali and Borneo. The dynasty had close relations with India. According to the Arab, Ibn Rosteh (903 AD), "The great king is called Maharajah. He is not regarded as the greatest among the kings of India, because he dwells in the islands . . ." The galleries of the stupa are covered with sculptures illustrating scenes from Buddhist texts. The sculptures are the finest examples of Indo-Javanese art, and the stupa stands as one of the great wonders of the world.

The kingdom of Cambodia rose to great power in the sixth century and lasted 900 years. Angkor Wat and numerous temples in Cambodia tell the story of their grandeur and magnificence. Angkor Wat is not a Buddhist monument; it is a shrine dedicated to the Hindu god, Vishnu, whose image stands on the top of the terraced structure. Now that Angkor Wat has been released from the deadly embrace of the rain forest it has reappeared as another wonder of the world.

A visit to the island of Bali provides another kind of experience. At the beginning of the sixteenth century Java turned to Islam and the Hindu royal family and a large proportion of the Hindu population took refuge in Bali. Since then Bali has remained Hindu with well-preserved Hindu customs and ways of life. Some foreign visitors in modern times searching for paradise on earth have found the answer in Bali. Indeed, the Balinese are among the finest people one can come across anywhere in the world. They enjoy a simple lifestyle and possess great artistic talents. Every home is a hive of industry with father, mother and children engaged in some handicraft. Painting, carving, artistic weaving — all seem to come naturally to them. Every village has an orchestra and all adults and older children take part in music, dance and folk plays. We saw one in the moonlight. The theme is nearly always from the Indian epics *Ramayana* and *Mahabharata*. We were told that every girl must take part in the dramatised lore to qualify for marriage. She is then regarded as being mature and adult because she has experienced in a living way the folk traditions of her people. The Hindu culture is still the basic culture of the people of Java though their religion is Islam. In every home I visited I saw paintings from the story of the Indian epics hanging on the walls.

Another fascinating chapter in India's history is the interchange between India and China following the spread of Buddhism from India around 200 BC. We read in books of history how later bands of Buddhist pilgrims came to India. The final stage of the perilous journey lay on the route originally taken by Alexander the Great, through Afghanistan, the Khyber Pass and Peshawar. This subject had always interested me. Twice I passed this way in the course of my travels in these countries. I saw evidence of this ancient cultural interchange in the Khyber Pass and in Afghanistan in the form of Buddhist stupas and images. The two colossal standing figures (Buddhist) carved on the face of the mountain in Bamian on the route to Afghanistan is one of the most arresting sights. I also wanted to see the route from the Chinese end, and while I was in China in 1948 I had asked for but was refused permission and facilities to go to Sinkiang. When I accompanied India's Vice-President, Dr. Radhakrishnan, to Kyoto in 1956 the University presented this most learned academic and myself with sets of the 29 volumes which depict in magnificent plates Buddhist art of 1500 years ago which was found in the Yun-Kang caves on the borders of Inner Mongolia. The caves apparently were a resting place for pilgrims travelling to and from India.

India's trade relations with the Mediterranean countries in the ancient days are well-preserved in history. In my constant travels in this region the shadows of the past were ever present. The sea route was from India to Alexandria via the Red Sea and travellers were then carried by land or by the canal system of the Nile. The land route lay through Persia along the shores of the Caspian Sea to Syria and Asia Minor. During the early centuries of the Christian era, Palmyra, Gerash and Petra were the principal centres of trade on this route. Pliny complains about large payments made every year by Rome for articles of luxury from India. In Rome, the piazza in front of the Arch of Janus next to the Circus Maximus was one of the important meeting places of the merchants of the East and the West. One wonders if the stone figure of the 'Bocca della Verità' still standing with open mouth next to the piazza was used to enforce fair trade!

The story of Palmyra came to life when we visited it during a Regional Conference held at Damascus. We flew to Palmyra from Damascus. The hotel we stayed in was named after the Palmyran Queen Zenobia of the third century AD. It was during the Roman times that Palmyra was transformed from a mere halting place to a city of importance. The trade from India, China and southern Arabia passed through it. It received the status of a Roman colony and Palmyrans became Roman citizens. In the Parthian wars it became prominent and was mistress of the Roman East. It was during these wars that the Palmyrans became restive for independence. Queen Zenobia took over the reins of government on behalf of her son, still a minor. She struck out boldly and captured Egypt. Emperor Aurelius would have none of it and led a campaign against her. He attacked her armies and carried off the Queen in chains to grace his triumph at Rome (272 AD). The city never recovered its splendour or importance after this downfall. We spent a day wandering among the ruins of what had once been a city of noble structures.

Gerash lies twenty miles north of Amman. Originally colonised by the veterans of Alexander the Great, it was rebuilt by the Romans, burned by the Jews, and desolated by Vespasian's captain, Annius, in 65 BC. It then came to be included in the Roman province of Syria and rapidly rose in importance and prosperity benefiting from the trade passing through. Magnificent buildings were built which, from the point of view of classical art, came to be regarded as triumphs of architecture. They lie in noble ruins now.

We approached Petra through a long narrow gorge with overhanging rock rising sheer on both sides into the sky. The gorge is too narrow for even a small car to pass. We were told that a few days before, a tourist party had been trapped in a downpour of freak rain and several people had perished. One of the arresting sights within the gorge is the pink-coloured facade of the Nabatean Treasury exquisitely carved out of the rock. Near the end of the gorge stand the most elaborate ruins, which in times past were hewn out of the multi-coloured rock special to the place, as well as Roman ruins of columns and structures. The walls of the mountain around the city are lined with rock-cut tombs of different types (Greek, Egyptian, Syrian), testifying to its position as a great trading centre.

I travelled widely in Egypt. Apart from the usual tourist sights along the Nile I went into the deserts to see many things not accessible to the normal visitor. I saw whole villages half-buried under the advancing desert sand and at the same time I witnessed the stand being taken to stop desertification by Egypt's Desert Control Authority. I saw some ghost towns abandoned for lack of water but whose monasteries remained almost intact.

One important journey I made was to the Siwa oasis 350 miles WSW of Cairo in the Libyan Desert. From Alexandria we travelled to Burg el Arab noted for its white sand beach where, as the story goes, Anthony landed to fulfill his dream of sensuous glory in the company of Cleopatra. From Burg el Arab we proceeded to Masra Matrûh. On the way we stopped to see the vast war cemeteries at El Alamein with the British graves on one side of the road and the German and Italian graves on the other. What a sad testimony to human folly! From Masra Matrûh we drove through the desert to reach Siwa. The Siwa oasis owes its fame to the Oracle Temple of Amon. When Pharaoh Akhnaton introduced the worship of One God Ato (the Solar Disc) putting an end to all sectarian temples including the great Ammon Ra, the Lord of Karnak, the priests of Ammon Ra decided to set up this Oracle Temple far out in the desert (c.1400 BC). The Oracle was still flourishing in 332 BC when Alexander the Great came to put down the rebellion in Thebes. But his preoccupation was the Persian Campaign and he came all the way across the desert to consult the Oracle and seek reassurance. I visited the Temple. It was in ruins but the walls were still standing, with inscriptions dating back to the fourth century BC. Near this temple there are scanty remains of another of the same age.

Having read the story of Athanael and Thaïs in the romance of Anatole France in my younger days I went in search of some of the old Christian monasteries to see how the monks lived. A visit was arranged. We flew to a Red Sea port in a small plane and from there we were taken in a motor convoy into the Sinai Desert, for over 200 miles. The journey was difficult, but when we arrived all the trouble seemed worthwhile. Surrounded by pink, harsh, bare hills, with only a patch of agricultural land for the monks to subsist on, the monastery looked the kind conceived by the founders — completely cut-off from the world, fully coenobitical in the regulation of the life of its monks. We were taken round the little museum where some interesting relics were preserved. But what struck me most in this monastic life was the way the monks disposed of their dead. They had only a tiny patch of space inside the monastery, no more than eight to twelve feet, to bury their dead. They would take out the body as soon as possible and only keep the skull to dry. I found a whole mound of skulls piled high beside this patch of land. We spent the night there. The journey back was eventful because we lost our way. Fortunately, the clouds opened up to show us the setting sun. We followed the direction of the sun to reach the Red Sea and then we were able to find our way.

I made use of my visit to Iraq to explore the historic past of this Mesopotamian region. I knew I would have no time to investigate the possible sites of Nineveh, Nimrud and Ashur. I concentrated on the ruins near Baghdad — Babylon being high on my list. Situated on the left bank of the Euphrates the ruins of Babylon that still remain are worth seeing. One mound, which for some reason still appeared to have escaped excavation, was pointed out to me as containing the palace where Alexander the Great at the age of only thirty-three died after a fever, with his dream of world unity still unrealized. On the walls of another mound, the sculptured figures of horses in relief was an arresting sight — the upper layers, we were told, had been removed by German archaeologists at the turn of the century for the Berlin Museum. The Tower of Babel did not appear to be genuine. Of the Hanging Gardens nothing remained except the bare slope of a hill. Ctesiphon was another place we went to which is famous for the ruins of the great vaulted Hall of the Sasanian period, one of the most magnificent structures of Mesopotamia. The great arch which was the first ever to be built still stands.

In another trip we were able to see some of the glories of Old Persia. Isfahan, which Shah Abbas the Great made into one of the largest and most beautiful cities of that time, remains almost intact. Its history goes back a thousand years. After crossing the picturesque bridge which has piers that date back to Sasanian times, we came to the Jewish Synagogue in which Jews said to have been settled by Nebuchadnezzar worshipped. The square in the middle of the city (the Maidan-i-Shah) was surrounded by mosques with domes and minarets exquisitely enamelled with faience-mosaic and painted tiles. Here you also find the former royal palace with courts and pavilions, one of which is famous for its verandah and throne room. The game of polo and gladiatorial combats were held in Maidan in the city's golden age.

Persepolis is well-known to the modern world. Planned by Darius the Great in the sixth century BC, Alexander in the course of his campaign burnt it down, justifying his action as 'the revengc of Greece for the burning of Athens by Xerxes'. The pillars of Darius' Audience Hall are some sixty feet high and thirteen of them are still standing and can be seen from a great distance. The capitals of the pillars in the form of animals are now lying on the ground. The rich carving of figures on the stone balustrades which depict the Medes and the Persians bearing gifts to the Shah or celebrating the New Year festival of the Vernal Equinox still proclaim the glory and grandeur of this great city. Not many years ago it was in Persepolis that Reza Shah Pahlavi gathered the Heads of State from all over the world to celebrate the 2500th anniversary of the Persian monarchy! Shiraz, 32 miles away, has the garden tombs of two of Persia's greatest poets — Sa'di and Hafiz.

Another ancient monarchy whose extinction I witnessed was that of Ethiopia. I had a special regard and admiration for Haile Selassie for the way he pleaded the cause of his country before the League of Nations. When I visited the country I was received with ceremony in his Throne Room with a leopard sitting on each side of his throne as a symbol of his kingly power. At the end of the audience when I turned to leave I was not altogether free from anxiety about the disinterestedness of the two leopards! One of the mementoes of my visit is a scroll on deerskin telling the salacious story of King Solomon and the Queen of Sheba pictorially in the traditional style. In the holocaust that followed the Emperor's downfall I lost some of my friends.

This is not the place to recount the tortuous history of the Crusades. Our Mediterranean Study devoted much attention to this region.* I visited most of the places named by the Crusaders: Sidon, Tyre, Jerusalem, Antioch, Aleppo, Emesa (Homs), Krak des Chevaliers, Tripoli, Acre. Once while travelling by road from Amman to Damascus I stopped at Krak des Chevaliers and spent a few hours exploring the fortress. The fort was built by the Knights of St John about the beginning of the thirteenth century. Earlier the great Saladin with the crusading spirit that the Latin Crusaders had lost had swept down on the adventurers and set the sacred Aksa mosque in Jerusalem free. The Crusades did serve a purpose by providing a meeting ground between the East and the West with incalculable long-term effects on both sides. But the long night of the people of this region is still far from over.

*See *The Mediterranean Project,* in Chapter 18, page 178.

Travels in Europe

I CAN TOUCH only very briefly on my travels in other parts of the world. In Europe both my wife and I were interested in art and architecture and we had plenty of opportunity to follow our interest. We were especially interested in Gothic architecture in France, Spain and England. In Spain the impact of 700 years of Moorish rule was everywhere in evidence — Granada, Seville and Cordoba in the south, and Toledo and Zamora in the north. Granada is resplendent with memories of the Moors. The story is told of the last Moorish King, Boabdil, in Townsend Miller's *The Castles and the Crown*:

> Boabdil rode on, away, out of history. As he climbed the mountains, he turned back for a last look at his ravishing city, the pride and alabaster eyrie of all his forebears, the rosy pearl that he had lost for ever. He could no longer contain his grief, he burst into tears, and Aixa (his mother) in her own bitterness said, "Weep, son, like a woman for what you could not defend like a man!"

Ferdinand and Isabella freed Spain from the long Moorish rule and are buried in the cathedral in Granada. Their life-size figures in marble decorate their tomb. What a personality Queen Isabella was! What courage, what single-mindedness in pursuing her objective!

We had a conference in Nîmes in the south of France to review the progress of the Mediterranean Project. Nîmes was one of the richest Roman outposts in ancient Gaul. The Temples of Diana and Apollo, the gateways, the Tour Magne, the aqueduct, the amphitheatre, the fortifications — all testify to its historic past. Carcassonne, a picturesque walled city of pre-Roman origin where we stayed for one night on our way to Spain, was one of the points from which the Crusaders embarked on their adventures.

In Germany, while passing through the Black Forest, we visited a Forestry School where the students turned out in their green uniform and plumed hats to sound bugles to welcome us reminding me of Robin Hood. We were met at one of the southern points of the Rhine by my friends, Minister Lübke and his wife who took us up the Rhine by boat to Cologne. Lübke became the President of West Germany soon after. Twice I visited West Berlin in connection with their 'Green Week'. This gave me an opportunity to compare Berlin as it was in my student days after the First World War and then, after the devastations of the Second World War.

By the time of my second visit the Berlin Wall had already gone up. I saw a small group of men and women with a child help aloft and I was told that the grandparents of the child in the building just across the Wall were anxious to see their grandchild. The scene reminded me of what I had seen in Jerusalem after the UN Mission had divided that city by drawing a line on the map.* I travelled widely in West Germany with my wife and family. We visited Berchtesgaden, Hitler's hide-out in the Bavarian mountains. Hitler certainly had an eye for natural beauty.

We travelled through the Scandanavian countries in one trip — Denmark, Finland, Sweden and Norway. In Finland I went to see the stadium where the 1952 Olympic Games were held. The statue of the Finnish national hero, Nurmi, one of the greatest runners of all time, stood at the entrance. An interesting feature in Finland was that every farmer had, in addition to his agricultural land, a small lake for fishing and a patch of forest for his fuel and other needs. In Sweden we went as far as Kiruna the great iron-ore mining centre. While crossing the Arctic Circle the plane made a dip to honour the event! In Norway, we went almost as far as the North Cape. On the way we passed Narvik which was made famous by naval battles during the last war. On another trip, I visited Iceland and Greenland. Iceland is a land of volcanoes and hot springs with the part of the country to the north covered in glaciers. We paid a visit to Geysir to see its world-famous geyser spouting its intermittent hot water and steam to the sky. This hot water is the main source of heating for houses and for vegetable cultivation in hot-houses. Even while we were there a new island was thrown up in the south by volcanic action. I also had an opportunity to watch whales being brought in from the sea by boats, one on each side to keep an even keel. I watched for a while and saw how such an enormous carcass could be reduced to nothing in a short time at the hands of expert cutters. What struck me most in Iceland was the character of its egalitarian society. Here there were no rich and no poor. Everyone had the necessities of life at a reasonable level. It was the most impressive society I have come across in my experience round the world. Democracy as a concept has undergone many changes since the Greek days but here was true democracy. I was taken to the spot, not far from Reykjavik, where a national Parliament (Aalthing) in the modern sense was first born. Some years before, Winston Churchill had inaugurated the 1000th anniversary of this event. Those were the days when fish disputes did not intervene between two democratic countries.

The more interesting trip was to Greenland. The legend goes that when the Vikings sighted 'Iceland', they called it this since they thought it could not be otherwise at that high latitude. But they had not taken the hot springs into account. So, when they sighted Greenland, they gave it a name which was optimistic but totally inaccurate. Until 1951 Greenland was closed to the outside world. No direct air service had developed even in the 1960s when I was in Iceland. I had to go back to Copenhagen to take the flight to Greenland. As we approached Greenland, the pilot announced that, owing to cloud-cover, he would

*See Page 126.

have to fly over to Nova Scotia. As I was nursing my disappointment the clouds suddenly opened and we were able to touch down, not at the international airport but at some obscure place down south. We took shelter at Narssarssuaq where we stayed for five days waiting for visibility. I used this time to visit places nearby in a small boat. We had to pick our way carefully through the ice floes. We saw some people whose boat had been crushed by ice floes and they had to be rescued by helicopter. After those five days when the provisions in the small inn had almost run out — all the supplies came from Denmark — the sky cleared and we were able to take off in a helicopter for Godthåb, the seat of the Governor of West Greenland. I stayed there several days to acquaint myself with the problems of this almost empty country. I was received by the Governor and some of his principal officers, invited to various homes and shown some of the sights including a small museum which showed how seal-hunting (which until recent years was the main occupation of Greenlanders) was practised with kayaks — the native Eskimo boat completely covered with seal skins, the covering being laced about the paddler so as to make him unsinkable. Dog-sledges were still in use but I had no opportunity to try one. I also attended a meeting where native Greenlanders from long distances had gathered to discuss their problems. It was clear that Denmark had a very liberal and enlightened policy towards Greenland. Their annual expenditure ran to hundreds of millions of dollars without commensurate material return. Their policy was to assimilate the Eskimos by education and intermarriage — a policy which characterises all Scandanavian countries — so unlike the policy of other countries dealing with an indigenous population. This vast land mass is covered with a thick sheet of ice miles deep in places and there is, of course, no possibility of agriculture. They find it hard even to bury their dead and sometimes graves have to be blasted from the ice with dynamite. Fishing now displaces seal-hunting and the Government is concentrating its resources on the south-western part of the coastland to develop the fishing and processing industry in the hope of attracting the small scattered settlements in the north and the east to this area. The east coast is exposed to very severe Arctic conditions and is almost uninhabitable. Only a few months before, a modern Danish ship, equipped for these conditions, was making its maiden voyage but disappeared in this area without leaving a trace. On our way back, we reached Søndre Strømfjord by helicopter and I stayed there for the night at an American Military Base. A photograph of the signpost at the airport gave the distance to the North Pole and other destinations in miles which I found interesting. I have included this in the Photograph Section, between Pages 214 and 215.

Before I leave Europe I must mention Greece. I visited all the usual tourist attractions both on the mainland and in the 'Isles of Greece', "where burning Sappho loved and sang". I stopped for one night at Olympia where the Olympiad started nearly 3000 years ago only to be discontinued in 394 AD by the Roman Emperor Theodosius. It started with only one item — a flat run and judging by the size of the course was probably not longer than 200 yards. But what was more important for the world was the philosophy behind it and from which we have so significantly departed. The games were to be strictly non-political in character giving access to all competitors coming from the warring Greek States. The competition was personal and the States did not become involved. There were no gold, silver or bronze medals, nor national anthems for the victor, only a simple laurel wreath to crown the head. When, in 1894, Baron Pierre de Coubertin called an international conference to revive the Olympiad, these basic principles must have been discussed. But as time passed, the rising tide of nationalism has swept these principles away. What we now see is a form of War not Sports, though as a role-playing substitute for real war it has its merits.

African Journeys

IN MY VISITS to the North African countries, apart from Egypt, I gave most attention to Morocco. My first visit was when the present King Hassan's father, Mohammad V, was alive. Since then our Mediterranean Study of the Rif and Sebou regions brought me into closer contact. At a later stage I paid my final visit to see the completion of the projects and King Hassan sent word round that a friend of the Moroccan people was coming. My wife and I were given a tumultuous welcome. In the Rif, the traditional riding and firing sport was organized. The banquet in the middle of the day under canvas was in the true Moroccan style with an entire roasted sheep on the table. We were given a right royal welcome at Fez where we stayed for a day or two, an interesting old city with its Suk (covered shops) and old Madrassas, University and mosques. Fez was regarded by the Arabs when they spread out over North Africa as the far end of the world. It has always been held in high esteem as a place of pilgrimage for Moslems. On the day we started from the Rif to go down to the south the road was crowded with ordinary country people, men, women and children — all dancing to music. When we reached our Rest House at the end of the day, a crowd had already gathered to meet us. I joined three other visiting dignitaries for tea. It was the party of Edward Heath, then Britain's Minister of State, who was resting from his labours having failed to secure Britain's entry into the Common Market at the Brussels meeting. He asked me why such a crowd had gathered. He expressed surprise when I explained that it was a reception party in my honour! Among my 52 photo albums, the one presented to me in Morocco depicting the scenes of my visit is a most treasured possession. Morocco, ever since its independence, has had difficult problems to face internally as well as externally. Moroccans are some of the most intellectually alive and warm-hearted people one can come across. Whenever I think of Morocco, apart from Fez, the picture of Marrakesh comes to my mind. The old city walls, men and women in their flowing hooded robes, the Atlas mountains dominating the scene — all are indelibly marked on my memory. I often visited the main square which is the traditional meeting place for people from the north and the south and it is a scene of dancing to Moorish music, wrestlers, snake-charmers and fortune-tellers, all busy displaying their talents.

Sudan was interesting in many ways. The British had a special brand of civil servants for this country — on the lines of rulers in *Plato's Republic* — and gave much attention to its agricultural development. When I was there the statue of Lord Kitchener still stood on its original pedestal but has since been removed. His finger was pointing towards the spot where some British soldiers were mowed down by the Al-Mahdi followers. The Grand Residence for the British Governor is now the Presidential Palace. It was in this Residence that General Gordon made his last stand. I saw the staircase from the balcony which he walked down to meet his death. When we met in the evening on the Palace lawn for the Presidential Banquet I could not get the shadows of the past out of my mind. One of the Ministers in attendance was, I understood, a descendant of Al-Mahdi (Seyyid Abdulla). He was to die later in a military coup which unseated his government.

In one of my visits to Tunisia, I stayed at Carthage. The classical illusions came flooding back: its foundation by Princess Dido and her love story with Aeneas immortalised in poetry and song; the famous campaigns of Hannibal; the "Delenda est Carthago" cry of the Romans, fearful of its prosperity. All this came to my mind. The thought that intrigued me was where did Hannibal get the thirty-seven elephants in 218 BC for his invasion of Italy across the Alps. It saddened me that the forests north of the Sahara which might have been their homeland have long since disappeared.

I deferred my visit to the parts of Africa south of the Sahara until the early 1960s because many of them were on the verge of celebrating independence from colonial rule. My first visit was to Ghana. Nkruma was then President. I was accommodated in the State Guest House which had been fitted out in the expectation that Accra would be the seat of the newly formed Organization of African Unity (OAU) with Nkruma playing a major role. It had regal splendour. Ghana's economy, largely based on cocoa, was also thriving. My visit to the West African States was combined with a Regional Conference. It was held in Abidjan in the Ivory Coast, a city which is regarded as the Paris of Africa. In its general layout the city does look impressive. When I got away from the cosmopolitan city into the countryside the atmosphere was quite different. The people still appeared to be following the old tribal customs and way of life. Indeed, this seemed to be the common feature of most of the African countries — the cities giving the impression that the people were very near the European standards while the masses remained outside the charmed circle. Countries like Kenya and Tanzania under their progressive leaders, however, appear to be breaking away from this general pattern. I must confess a special love for these two countries because of the Wildlife Reserves they maintain which they inherited from the British days. When I went to Nairobi to attend the Regional Conference, the first thing I did was to visit the Reserve which was near the city. I am afraid I neglected to call on the Minister of Agriculture, who was the host. When the Conference opened the next day the Minister in his formal address expressed his grievance that I seemed to be taking more interest in animals than in human Problems. I did not try to contradict him as his criticism was at least partially correct.

I enjoyed a night at the famed Tree Top Hotel and was guided there by the 'White Hunters', a corps which gives protection to dignitaries against wild animals. I stayed up all night watching the animals coming in herds for the salt that the earth round the lake provided. It was an unforgettable sight. I saw the elephants with the big tusker guarding the herd, moving round and round, ever vigilant. I saw the mother elephants suckling their babies. The sight of rhinoceros couples that fought and grunted all night just a few yards away was certainly not an invitation to sleep and happy dreams. On the way back, I was taken to see the flamingoes covering the lake in great flocks. These birds are becoming scarce in other parts of the world, in eastern India in particular, and I was fascinated by the profusion here. From Kenya I flew on to the Serengeti Reserve in Tanzania in a small plane piloted by a British gamekeeper. We flew very low and saw the movement of the animals. I spent the night at Serengeti. I was accommodated in the hut put up the year before for the visit of Prince Philip. My two African colleagues were settled in tents. We were warned that lions were around the area in numbers and that we should be careful. This was probably said in jest but having seen some lions prowling around during the day, we were not fully reassured. At night I woke up to the creaking noise of the front door. I sat up to face the lion and an uncertain end but the door had only swung open in a rising wind! Next morning I learnt that my African friends had stayed up all night on guard! We went round the Reserve again in the morning and when we returned there was a lion sitting licking his paws and growling right outside the door of my hut! My visit to Uganda was uneventful. Many Indians lived in the country engaged in business. Many of their forefathers had come to work on the construction of the railway by the British rulers. I saw the spot where Speke first discovered the source of the Nile and a marble slab to commemorate the event stands there.

I made a special trip to what was once French Africa but has now been dissolved into many States. My first stop was the Central African State. When President Bokasa received me he had a French 'Advisor' in full military uniform sitting beside him. Only a hundred yards away there was an outpost of the French Foreign Legion manned by several hundred Legionnaires.

Chad provided a different experience. I went there to see an irrigation project which was drawing water from one of the Chad lakes. These lakes are known to have been the source of the Nile in prehistoric times. I remember that we had to walk quite a distance without any head cover and were exposed to the sun with a temperature into the hundreds. We passed through a village where two American Peace Corps volunteers were living. I admired their idealism and tenacity. The villagers who worked on the project were carrying earth for a dam in small cooking bowls or baskets. A heavier load was not possible for them.

I passed through other countries. I was left with the impression that while granting independence the French had not loosened their hold on trade and industry. Perhaps they succeeded in doing this by the way they had ruled their colonies. French culture had been transmitted through the system of education the French introduced and they had not kept themselves aloof socially from the local population in the way the British had done. In this way they created an élite class with French ideals and French culture who appreciated their close association with the French expatriates. All this was facilitated by the persistence of the basic social structures in the countries with tribal loyalties and customs insulated from the élite class. There was also no comparable traditional culture to inhibit the élite from absorbing French culture freely. At most of the African ports, I noticed that the ships waiting to be loaded or unloaded were all French. At one port, I saw logs floating for miles in the sea awaiting French ships to transport them away. Under the new régime, the French were keeping a low political profile but there was no mistaking their real interests were being well guarded. In international fora, the French delegates always proclaimed their generosity in their policy of assistance to the developing countries — more than 1 per cent GNP as against the 0.7 per cent recommended by the UN General Assembly. One would have to realize that most of the French assistance went to their former colonies. The larger proportion of EEC assistance to Africa in the early years was also I believe, at the behest of France. Not that these countries did not need the assistance but the background to the assistance is interesting. This background may also explain to some extent the all-too-ready acceptance by African countries of the recent military interventions by France. However, things appear to be changing now.

Expeditions in South America

I TRAVELLED extensively in South America. Once before, I had travelled up the Amazon which gave me a general idea of the vast hinterland of Brazil especially the forest region. In recent years Brazil has embarked on bold measures to open up the interior. The vast forests of Brazil have an environmental importance for the whole of that continent and beyond. This needed to be emphasised and Brazil is becoming more aware of the dangers in overexploiting her forest wealth by financiers and industrialists from abroad. Argentina presented another kind of picture. To begin with I had difficulty in establishing my identity with the President who had just taken over after a military coup. The agricultural resources of Argentina, if properly utilised, could make it one of the richest countries in Latin America. Only a few years ago Argentina was a major exporter of wheat and meat. Today she has one of the highest rates of inflation in the world.

I was fascinated by Chile. The intellectual class impressed me with their progressive ideas. With its rich mineral resources and its intellectual élite, Chile could be one of the leading countries of Latin America. It had inherited a culture with constant fertilisation in the past centuries from contacts with British and French societies in particular. Valparaiso in pre-Panama-Canal days used to be one of the most important ports of call for ships from Europe. It was the Valparaiso coast that 150 years ago welcomed the *SS Beagle* which had Charles Darwin on board, busy with his explorations on the Origins of Species. My last visit to Chile was in 1966 when there was not a speck of cloud on the political horizon. I had planned to go down to the furthest point of Chile which is Punta Arenas to see the ice and snows of Antarctica and the stormy seas around the Cape. I was disappointed when I could not make it. While in Chile I saw the massive figures carved out of volcanic rock. Similar figures are also to be found on Easter Island, 2000 miles across the Pacific. This had led to the theory that at one time in the past there was migration of people from the western coast of South America to the Pacific islands in boats made of balsam logs. It was to test this theory that Thor Heyerdahl in 1947 undertook his historic *Kon Tiki* voyage.

I saw the remains of the lost civilisations of the Aztecs and the Mayas and was in the heartland of the Incas. After I had been photographed at Lima Airport with the freshly proclaimed World Beauty Queen of that year I took the plane to La Paz (Bolivia). La Paz stands at 13,000 feet above sea-level. I remember how I reeled to the Waiting Room gasping for breath. And every Minister I called on that day seemed to have his office on the second floor and I had to pant all the way up the steep staircases. Next morning, when I recovered I took a boat trip around Titicaca Lake which competes in height with the great Manasarowar near Mt

Kailas deep in the Himalayas, the abode of the Gods according to the Hindu legends. I saw boats made of reeds which the Indians traditionally used. I also saw the hovels in which they lived and witnessed their abject poverty. In the centre of the city itself I passed through the market where the Indians were allowed to sell contraband goods as a concession by the Government. There is a glaring contrast between the rich and the poor. Though the Patiños and the Hochschields and the Aramayos, with their multi-million dollar fortunes from the tin mines, are no more, the mass poverty of the Indians who form two-thirds of the population remains as before. The nationalisation of the tin mines was hailed as promising a new era. The expectations have not been fulfilled. I had arrived at La Paz just after one of those endemic military coups and I could hear shooting still going on at night. In the midst of this melancholy situation, the sight of that rich cluster of eucalyptus trees near the airport still lingers in my mind. I wondered at the tenacity of this species to survive and flourish at that height and in that barren land.

I have related some of my diplomatic experiences in the USA in Chapters 5 and 7. I also travelled Canada from end to end, though I did not get to Hudson Bay — I had very much wanted to see the living habits of the Eskimos in that northernmost inhabited region. In both these countries, immigration has become an increasingly difficult problem. Statistics tell us that over 40 million immigrants had come to the USA mainly from Europe by 1930. The outflow of British immigrants to the USA, Canada, Australia, New Zealand and Africa since 1821 has run into many millions. Ironically, the British, who had recruited over one and a half million Indians as indentured labour for building railways and running sugar plantations in their colonies, now suffer extreme vexation at what they see as a threat to their national identity at the entry of a few hundred thousand ethnic Indians, many of them descended from that same indentured labour.

It is debated these days whether the Westminster Parliamentary-type of government or the American Presidential-type is best suited to the conditions of newly independent countries. The relations between the President and the Congress in the United States, under the principle of checks and balances, has always been delicate. The classic example was the repudiation by the Senate in 1919 of the Covenant of the League of Nations of which President Wilson was both the inspirer and its main architect. How this single act changed the whole course of history! In recent years, from time to time, the imbalance has become even more manifest.

There is, however, another model of the Presidential system which calls for notice — the Mexican. Under the Mexican system, the monolithic majority party, which has a history of its own, forms the keystone and its most distinguishing feature. The party chooses the President though the announcement of the nomination is the prerogative of the incumbent President. The nomination is then put to direct popular vote which in the circumstances becomes no more than a formality. Having thus elected the President, the majority party then assures the President of the support that he may need in both Houses of the Congress for the almost unlimited executive and legislative powers he enjoys under the Constitution during his non-renewable six-year term of office. There is thus no room for conflict between the President and the Congress as in the United States.

There is still a third Presidential system, which I would hesitate to call a 'model' in this context — the French system evolved by de Gaulle in the special circumstances of post-war France. The President is elected by general suffrage and not by an electoral college. He is to have personal control of government policy. On critical questions, he can seek renewal of confidence by recourse to referendum. The system is based on the idea that power rests on popular control and not on a system of 'political parties' in Parliament. The system, however, retains certain essential Parliamentary functions such as the right to criticise the government or individual Ministers and to withdraw confidence by majority vote, requiring a General Election. This system of ruling by referenda, while appropriate in the conditions of the tiny old Greek city states, would seem to be hardly transplantable to the complex conditions of modern times except perhaps to reflect the personality and genius of a figure like de Gaulle.

As I am writing, the news of the assassination of President Pak Chung Hee by his own 'Brutus' has just come through. This brings many memories and thoughts. My first acquaintance with the complex problem of Korea came, as I have described in Chapter 8, when the UN debated the post-war Korean Question (1949). I came to know Pak Chung Hee soon after he had taken over by military coup, and my contact with him continued all through his eighteen years of office. We came to like each other. He certainly was a dictator, but I always thought that some of the opprobium he acquired as a dictator was due to his natural shyness and his inability to communicate his thinking to those who opposed him politically. During his régime, particularly the second half of it, his country made spectacular economic progress. South Korea and Taiwan could be said to be the only two countries in the developing world which experienced an 'economic miracle'. There was a certain dichotomy in his approach to the problems of his country. Uppermost in his mind was his anxiety for national security. His argument was that there was a constant threat from the North and the country must pay the price by eternal vigilance. He would make no compromise on this point. But in economic matters he was prepared to give full freedom to private enterprise so long as it kept clear of involvements prejudicing the political and military stance of the country. He involved himself in working for the betterment of the conditions of the rural community. He inaugurated 'The New Community Movement' — and I paid a special visit in 1977 to study this movement. I found it to be one of the most successful rural development projects I had seen anywhere. But even this movement had to be within the framework of what he conceived to be the need for national security. The village leaders selected for training had to be chosen by the Prefects. Pak was a simple man who never sought material benefits for himself through his office. Perhaps he answered Churchill's description — "Dictators ride to and fro upon tigers which they cannot dismount, and the tigers are getting hungrier."

The world today makes the *word* 'Democracy' a god to worship. This god has now taken strange shapes and forms. Two hundred years ago, the great historian Edward Gibbon made the prescient remark that "Corruption is the most infallible symptom of constitutional liberty." One wonders what Gibbon would have to say had he lived to see the almost pervasive corruption that holds the world 'democracies' in its grip and what he would have thought of a 'dictator' like Pak Chung Hee.... "Watchman, what of the night ...?"

Chapter 25

Freedom From Hunger Campaign: The Later Phase

Review of the Campaign

THE CHORUS of the ancient Greek Drama could look both before and after and transport the audience far into the future. One did not have to evoke the spirit of the Greek Chorus to anticipate the course the Freedom from Hunger Campaign (FFHC) would take after I left the scene. The pressures I had to contend with all through the Campaign came from different sources: the pull of some of the developed countries and their NGOs interested primarily in fund-raising and their desire to have done with FFHC so that they could move on to another campaign to enthuse their contributors; the unrelenting determination of some developed countries to rein in the Specialised Agencies whose budgets were growing, particularly FAO with its FFHC; the protagonists of the UN Development Decade who wanted to be the sole actors in their theatre; the increasing trend in the developing countries to bring the national FFHC Committees more and more under their bureaucratic control using the NGOs as and when the bureaucrats thought fit; and others who became frankly jealous of the position that FAO had attained in world opinion through the successful operation of the FFHC.

These pressures I could resist because from the very beginning I made FFHC the central theme of all FAO activities. I interpreted the phrase "rededication to the objectives of the Charter" as requiring all FAO work, ongoing or new, at the centre or in the field, to be attuned to that theme. The normal work of the Technical Divisions was given a new orientation.[1] A World Soil Map was no longer to be regarded as one only for the specialists working on correlated soil cartography, but one through which the newly independent countries could develop their national plans for extended agricultural activities. In forestry, as the successive World Forestry Congresses showed, the concern of the FAO Member Countries in 'scientific utilisation' of forest products was fully respected while at the same time respecting its 'conservation' role in controlling climate and the flow of water in its rivers. Under the impact of the Campaign, watershed management on the one hand, and rational use of forest products for community development on the other, assumed special significance. Fish are the main source of protein for many countries of the world and FAO's practical approach to fishery problems have been detailed already. At the first stage we devoted our attention to transforming small-scale backward fishery operations into larger modern commercial enterprises and to the creation of entirely new fishery industries. We then stepped up our work on stock assessment and survey of inland fishery resources. It was at a FAO Council meeting (1966) that the revolutionary idea of an 'Exclusive Economic Zone' (EEZ) of 200 miles was first debated as a world rather than as a regional problem. FAO is now playing a leading part in giving assistance to many of the coastal states, both developed and developing, to make EEZs work without serious dislocation of the supply of fish for world consumption. As for nutrition, it is a specific FAO Charter responsibility and though FAO had done much pioneering work in the past it was under the FFHC that nutrition found its proper place in FAO priorities. Rightly, FAO is now the Lead Agency in the UN system. Finally, the World Conference on Agrarian Reform and Rural Development (WCARRD) must be regarded as one of the most important results of the FFHC which paid particular attention at all times to involving the rural masses in the process of self-development.

So much for the orientation of organizational activities of FAO. All the major new initiatives undertaken by FAO during my time were rooted in the philosophy of the Campaign.[2] We were able to lay to rest the long-standing problem of surpluses by setting up the World Food Programme. The Indicative World Plan came out of the Declaration of the World Food Congress of 1963, and as I have explained, pulled together all the diverse activities of FAO into one combined operation. We were able to link up with the World Bank in a meaningful way, and later with the Regional Banks, to give a special impetus to development activities. Our contribution in the field of Commodity trade, which remains the bulk of the export trade of the developing countries, was made more purposeful by what we had learnt of the prevailing conditions in the developing countries in the field work associated with the

[1]For details, see *FAO and Field Programmes,* Chapter 22.
[2]For details, see Chapters 18 to 21.

Freedom from Hunger Campaign. Our link-up with multinational industries was due to our anxiety to accelerate the rate of development by setting priorities agreed between us and the governments concerned for the investment of their immense capital and their skills which normally would go only to the most profitable areas. In all this, I had made myself personally responsible for guidance and direction. This was made clear to all, and the pressures which built up from time to time could thus be more easily contained.

Later Developments

SINCE THEN I have been able to watch the developments of the Campaign only from a distance. I have before me the mid-term Review (1975) and the Expert Consultation Report on which it is based.

In 1971, the FFHC World Conference held in Rome showed up the old pressures. The Conference proposed a new orientation for the Campaign and a new sub-title. The orientation was to diffuse the focus of the Campaign to cover all developmental issues, on the grounds that 'a developmental crisis exists' (whatever esoteric meaning that may have for the initiated). In order to reflect that reorientation, they decided that the Campaign in future should be called FFHC/AD (AD standing for 'Action for Development'). At last the UN Development protagonists of 1964 had got their feet under the door!* This proposal for reorientation and the sub-title also satisfied those fund-raising bodies in the developed countries who could not keep up the enthusiasm of their supporters for any one campaign, however worthy. Another recommendation of this Conference was that the 'FFHC should have no field operational activities'. It is not made clear why, but if one reads it with their observation that 'FFHC can have an impact upon the policy and practice of FAO', it would seem that they looked upon FFHC not as an important factor in the conduct of FAO as a whole, but as something apart.

The FAO Council Committee of the Whole, which met soon after, apparently endorsed these suggestions. The reorientation and the sub-title were given effect to. The Committee made another observation in their report: that the Campaign phase of FFHC, focussing on information and public relations activities, was largely over. It is difficult to understand this observation. The heart of the Campaign as originally conceived and enshrined in the FAO Resolution of 1959 was to heighten awareness in the world to the depth and extent of hunger that existed and thus improve the climate for effective sustained accelerated action.

*See *Sequel to the Freedom from Hunger Campaign*, Chapter 16.

Without this 'information/public relations' content there could be no campaign in the proper sense of the word. This educational aspect was never intended to be a once-off operation. Each coming generation would have to be infused with the spirit of the Campaign until a stage was reached when the problem was well under control. We know that even at the end of the last and beginning of this century, hunger and poverty existed in Europe, particularly in the Scandinavian countries and Ireland and today, due to progressive social, educational and economic measures, they have been able to overcome this problem.

Here I would like to raise once again the old question — the wisdom of merging the FFHC into the UN's Development Decade. UN Development Decade is no more than an expression of hopes, laying down some attainable targets of economic growth and for all concerned to make a real effort. It is not a Campaign, in the sense of an organized worldwide action-programme over as long a period of time as necessary to attain the objective. It thus cannot be a substitute for FFHC which was intended to do just that. It is to the credit of the Council Committee that they insisted that the focus on agriculture and food be maintained. The 'scenario' as presented by the FFHC Conference, giving a new orientation, however remained, later endorsed generally by the FAO Conference.

One important element of this original campaign has, I am glad to see, survived this new orientation — the people-to-people movement, people's participation in their own development. This was a central theme of the Campaign all along and time has only strengthened it. All the examinations and reviews were unanimous that this element of FFHC was something which needed to be perpetuated. The Expert Consultation (1975) noted the following features of FFHC/AD: that it provided an international framework for operational collaboration between the governments and the extra-governmental sectors — especially in rural development and that it strengthened the capacities and potential for development action of the extra-governmental sector; that it offered an international focus, basis and framework for development education; and that it had proved itself capable of responding swiftly and effectively to contemporary issues.

People's participation for self-development as the very essence of the 'global development crisis' has now come to be accepted as a 'slogan' by all. The resolutions of the ILO Conference fully reflected this. The 1979 World Conference on Agrarian Reform and Rural Development (organized by FAO and UN) also had people's participation at the core of its approach to Rural Development.

So the *élan* has gone out of the FFHC and yet what remains can still be of infinite value. Will People's Participation prove to be the 'gene' passed on by FFHC? The World Food Conference (1974) resolution on *World Food Security* ("recognising the urgent need to ensure availability at all times of adequate world supplies of the basic foodstuffs . . . to avoid acute food shortage ... in the event of widespread crop failure ..."), followed by the formal adoption of the *International Undertaking on World Food Security* and, in 1980, of a revised *Food Aid Convention* with the objective of securing, through a joint effort by the international community, the World Food Conference target of at least ten million tons of grain as food aid a year — all have been significant manifestations of continued international concern on the same theme. The initiative of the present Director-General of FAO, Dr. Edouard Saouma, to underpin the concept of *World Food Security* by a 'five-point plan' with a view to helping the developing countries to cover their food deficits and build their own national reserves (as also his attempt to place the International Emergency Food Reserve of 500,000 tons a year on the same legally binding footing as the Food Aid Convention), demonstrates FAO's steadfastness in pursuing its ultimate objective as laid down in the Preamble to its Charter 'ensuring humanity's freedom from hunger'.

Chapter 26

End of the Day

A Farewell to FAO

I HELD CHARGE OF FAO for eleven years completing half its lifetime since its foundation. I think of this period as the most rewarding in my life. The Freedom from Hunger Campaign was my central preoccupation. To this international crusade, I devoted all my energies.

In his autobiography, Edward Gibbon describes the thoughts and emotions he had after laying down his pen, writing the last lines of the last pages of his monumental history: first, his emotions of joy recovering his freedom at long last and then, his melancholy having taken leave of an old and agreeable friend. Like the great, sometimes even the humblest may experience similar emotions. As I come to the close, my melancholy comes from the thought how far we still are from solving the problem of world hunger.

Yet we could claim some advance. The world today is conscious as never before of the problem of hunger afflicting millions of our fellow beings. No longer can any responsible government turn a blind eye to it without seeking its own destruction. At last, the vast masses are awake and their first demand is for food. Nor can the International Community take shelter behind the euphemism of malnutrition. The Brandt Report and the Cancún Summit have brought out the stark realities and placed the problem as the top priority for national and international action.

Another advance was in bringing out the relationship between population growth and hunger — the social and political implications that the rising rate of population growth presented. Until then, the population problem was regarded mainly as a matter for statistical study. The professional demographers in the UN Population Commission did not regard it as their responsibility to raise such issues, while the UN General Assembly kept itself at a distance for fear of offending religious susceptibilities over population control.

These could be claimed as our conceptual gains. What about the practical measures taken? The creation of the World Food Programme was one of our prime initiatives. This Programme laid to rest the problem of surplus disposal which had been a growing concern for some of the main grain-producing countries ever since the last World War. It showed a meaningful way to use food surpluses for economic development.

The link we established with the multinational financial institutions like the World Bank and Regional Banks has proved to be of major significance. Through the FAO Investment Centre now flows not millions but billions of dollars-worth of projects every year representing the cooperative efforts of national governments, the World and Regional Banks and FAO.

The Indicative World Plan for Agricultural Development (IWP) was a pioneering exercise in perspective study to assist national and international efforts in long-term planning. It is my belief that this study will provide a standard of achievement not only for FAO but for the entire UN family for years to come. It was the only detailed study of long-term future perspectives available when the United Nations International Development Strategy for the Second Development Decade (DD2) was prepared and adopted. The latest exercise undertaken by UN for a study of *The Future of the World Economy* has acknowledged its debt to our pioneering perspective study.

An important feature of the Freedom from Hunger Campaign was the working relationship we were able to establish with the Non-Governmental Organizations, national and international. We realized from the beginning that the solution to the problem of hunger was not only a matter of increased production and productivity. Statistical self-sufficiency did not necessarily solve the problem; the poorest sections of the population with little or no purchasing power could still remain hungry. We set out with the idea that equitable distribution was primarily a problem of modernisation of socio-economic structures which had remained stagnant for centuries. We felt it of the utmost importance that the rural people, who were the producers of food and who suffered hunger most, must be reached directly as far as possible. However, a peoples-to-peoples movement could not be undertaken through government agencies alone; for this we must have the active participation of NGOs. We were thus able to bring out the NGOs from their hitherto confined "Consultative Status" in the UN system which permitted them, apart from raising funds for worthwhile projects, to be in a position of somewhat distant and detached observers of the world food problems. Here I might add that I never believed in any magic year by which the problem of hunger could be solved. It called for a total war, to be continued, if necessary, for generations to achieve the final victory.

These were the practical gains. Where we failed was in our inability to move the big powers — the signatories of the Test Ban Treaty — to take a fresh look at the arms race and try and set apart a portion of the savings for the creation of a development fund. The arms race today has assumed menacing proportions. In the 1960s, powerful voices were raised against this appalling waste. Today such voices are muted. On the contrary, the arms industry (nuclear as well as conventional) is being openly encouraged by some countries, not only to serve their geo-political concepts but also to support their economies in recession. Presidents and Prime Ministers, not to speak of the lesser breeds, do not find it wrong to take upon themselves the responsibility of openly serving as selling agents of sophisticated and costly weapons of war. A strange schizophrenia has descended upon us. We accept a daily diet of news about development of ever-more destructive nuclear weapons, without feeling disturbed in going about our daily business. The story of little Peterkin as told by the Poet comes to mind. When little Peterkin asked his grandfather, "What good came out of it at last?", his grandfather could only reply, "Why, that I cannot tell, but 'twas a famous victory!" We do not seem to comprehend that this time there will be no victors. It will be full-scale nuclear war for fear of pre-emptive strikes from either side, sparing no one to tell the tale. Presidents, Prime Ministers, Comrades, Generals alike with the total population will be engulfed in the holocaust.

"To save succeeding generations from the scourge of war ... to promote social progress and better standards of life in larger freedom". That is how we started. Is it too late to recapture that dream?

Memories crowd upon the mind. Immediately after the decision to move the Seat of FAO to Rome, I came to Rome on transfer. And I recollect being consulted about the two alternative sites offered by the Italian Government for FAO. One was the Palazzo della Farnesina, now the Italian Foreign Office, and the other the marble palace built by Mussolini for his Ministero dell'Africa Orientale. In spite of its past associations, my clear preference was for the latter. My reason was that unlike the Farnesina, though it had more spacious accommodation and grounds, Mussolini's palace was right at the heart of ancient Rome — surrounded by three of the seven hills (Celio, Palatino and Aventino) and situated between the Baths of Caracalla on the one side and the Circus Maximus (the sunken arena where the chariot races used to be held) on the other, with the Arch of Constantine and the Colosseum right in front. The Foro Romano was just hidden by the Palatine Hill. The disadvantages of the site were of course clear. One side of the rectangular building would still continue to be occupied by the Italian Posts and Telegraphs. But I do not think FAO ever regretted the decision.

From my room on the corner of the fourth floor, I had a full view of all the sights. From my seat, I could watch the Palatine in the changing light of the day. Several times, I had to resist being shifted to some more spacious accommodation which was considered suitable for the Director-General. On my last day at FAO I clearly remember the sun setting on the Palatine Hill and bathing it in amber light harmonising with the amber of the ancient ruins. I watched a lone tree gradually disappear into darkness. In some mysterious way the whole changing scene seemed to fuse into my mood and I felt at peace in saying my last farewell.

Was all this labour in which millions joined worthwhile? We must leave it to history to judge.* But for me, the experience was enriching beyond words. It broadened my understanding of some of the basic human problems all over the world — the problems of poverty and hunger, the deprivations, the sufferings.

As I look back, I remember two speeches at my Farewell session of the FAO Conference:

> *E. Bourg de of Chad* — "Several months ago, Mr. Sen came to Chad. Throughout an extremely hot and uncomfortable day, I was able to gain the measure of his physical and intellectual resistance as well as his humanity. There was so much kindness in his eyes as he watched the men and women of my country building a dam and carrying baskets of sand on their heads beneath a burning sun. . . ."
>
> *J. L. Peñalver of Venezuela* — "The people and Government of my country pay their affectionate tribute to the great architect of FAO. The peasant families that now have land, a roof, food and freedom are the best monument that can be raised to a man, to an idea . . ."

On such an occasion, one must make allowance for overtones of emotion, Still I found their words deeply moving, making whatever little I have been able to accomplish worthwhile. And one can also turn to the Poet for comfort:

> "Ah, but a man's reach should exceed his grasp,
> Or, what's a heaven for! . . . . . ."

*See Appendix IX on Page 331.

Chapter 27

Postscript

My visit to the USSR

WHEN I RETIRED at the end of 1967 I left a busy and hectic life behind and I looked forward to a period of detached contemplation on life's experiences. But I soon found I could not rest. My life has continued as before, travelling, visiting friends, acquiring new experiences, albeit in a minor key.

The organization of the second World Food Congress, which I had planned as a sequence to the first one held in Washington in 1963, was taken over by my successor and was held at The Hague in 1970. I was among the three Special Guests invited by the Netherlands Government who were hosting the Congress. The other two were Mike Pearson, ex-Prime Minister of Canada and a founder member of FAO, and Lord Boyd-Orr, FAO's first Director-General who had also helped to give shape to this world organization. The Congress at The Hague fell far short of expectations and, except for some general resolutions, achieved little in real terms. None of the Special Guests had any part to play except that we gave preliminary addresses.

However, I was able to make some contacts which proved useful. I was invited to attend the Conference of Agricultural Economists which was to take place at Minsk (USSR) originated by Dr. Elmhurst of the United States. What excited me was not the Conference itself but the opportunity to visit the Soviet Union. The Soviet Union was one of the signatories to the FAO Charter at the Quebec Conference in 1945 but later refused to ratify it. Stalin had started his policy of revolutionizing agriculture in the USSR by abolishing private ownership of land and operating through Central Tractor Stations. In the process, millions of landholders (Kulaks) had been displaced and millions had died of famine. The loss of cattle no longer required for cultivation was also to be counted in millions. Apparently, this was not a scene to be exposed to international scrutiny. That was probably why the USSR did not ratify the Charter. The USSR has not revised this attitude since. Agriculture still remains the Soviet Achilles heel. The USSR has been the biggest buyer of wheat in the international market and this has become a painful routine every few years. However, this has not prevented the USSR from participating in important international conferences organized by FAO. At the 1966 Land Reform Conference held in Rome by FAO, the USSR delegation was headed by an eminent agricultural scientist who asked me why the USSR was not a member of FAO! The impression we had in FAO throughout was that while agricultural scientists and technicians in the USSR were keen on joining FAO, the policymakers — the Politburo — were disinclined to change the course already set.

The Minsk Conference was really nothing more than a fraternal get-together. No striking ideas, no breaking of new ground. However, since my main object in coming was to see something of the country this did not matter. There was little time or opportunity for any of us to make any independent study. We were all rigidly bound, by lack of language and facility of movement, to the *Inturist* Desk. But, fortunately, official sight-seeing tours were arranged for groups willing to see the country. My wife was with me and we chose the trip across Siberia to Irkutsk and Lake Baikal. Siberia always held a fascination for me. From my boyhood days I had visions of dog-sledges racing across the snow and ice of Siberia, with packs of hungry wolves attacking them and being kept at bay by fur-covered sharp-shooters! Since my retirement I also had had more time to delve into Russian history. That was why we chose Siberia in preference to Tashkent and Samarkand which are very much a part of Indian history and would otherwise have been our choice.

Our first stop from Moscow was Novissibirsk, the capital of Eastern Siberia. It was a night's journey, which in the sledge-days must have taken months. I would have preferred the Trans-Siberian Railway and seen something of the landscape, and perhaps even taken a sledge ride for a short distance just to see how it felt. But there was no time and a sledge ride was a pipe-dream. Novissibirsk is the largest town in Siberia and, apart from being a big engineering centre, it has a University and a Branch of the Academy of Sciences. It is the intellectual centre of Siberia. We stopped for a day and were taken to the Museum. The Director spread out on a table before us the riches of Siberia in diamonds, gold and other precious metals. He pointed out on a wall-map the large finds of petroleum and gas which were waiting to be exploited. The whole place had the atmosphere of a University Campus.

Some kind of exhibition was about to open and we found a lone Englishman busy tidying up his small inconspicuous stall. I felt rather sorry for him, all alone and very far away from his headquarters in London.

Our next stop was Irkutsk. It was difficult to imagine that Irkutsk was on the same longitude as Vietnam; we had come that far in so short a time thanks to the global shape of the earth. This used to be the final destination for the exiled Russian and Polish nobles who survived the arduous and hazardous journey across Siberia. It looked a very sophisticated town with a Cathedral, churches, museums and old timbered houses with exterior carvings and paintings. The Cathedral was a reminder of the past glory of Russia. We went in; everybody had to stand for there were no seats. The service was, of course, Greek Orthodox. A family friend in Rome — a Russian emigrée — was baptised there and I had promised that I would visit the church. The museum we visited was once the residence of a wealthy Armenian family now in exile or extinct. Everywhere we saw the imprint of the exiles from Russia who had tried to recreate their home atmosphere in this faraway place.

We spent a day at Lake Baikal which is the largest freshwater lake in Eurasia and the deepest in the world. They say the water reaches a depth of 132 feet. The most interesting feature was the unusual marine fauna and flora, including the Baikal seal. We saw some of these in the Museum. At Irkutsk, the thought occurred to me that we might have tried to include Outer Mongolia (Ulan Bator) in this trip, but the excessive red-tape in the USSR would not allow it. I also cast wistful eyes to the north and east of Irkutsk; there was so much to see and so little time to do it. We had an interesting ride over the lake in a hydroplane.

On our return to Moscow, we went on the usual touristic forays: the Kremlin attracts visitors from all over the world; the Uspenski Cathedral (1475-79) where the Tsars were crowned; the Archangel Cathedral (1505-09) where the Tsars were buried; the Tsar Kolokol, the largest bell in the world which was never rung, now placed on a pedestal after its fall from its scaffolding in 1739; the Tsar Pushka Canon weighing 40 tons which was never fired; the Cathedral Vasilii Blazhenry (St. Basil the Beatified) to commemorate victory over Kazan, dominating the Krasnaya Bloshchad (the Beautiful Square, more commonly known as Red Square) — we saw all of these. What impressed me was the care with which the Tsarist treasures were being preserved and renovated (after the damage during the War). The Oruzheinaya (Armoury) Palace has the richest collection of treasures. We also went to the Bolshoi Theatre to see the ballet *Giselle*.

The city itself has the appearance of a very new city, full of skyscrapers, Stalinist-style. Very little thought was given to the interior of these skyscrapers. At the Ukrainia Hotel where we were staying, there was no toilet on the ground floor which housed all the restaurants. One had to take an elevator and then stand in a long queue.

Leningrad was another matter. It was built by Peter the Great in 1707 as his 'window on Europe' and the capital was transferred there in 1712 from Moscow. It still has the appearance of a Western city. There is so much to see here and so much to remember of its past history. Zimni Dvorets (The Imperial Winter Palace), work of the Italian Rastelli, which was stormed by workers and sailors on 25 October 1917 to launch the Bolshevik Revolution, stands as it was and is visited by millions of tourists every year. The interior still brings back the pomp and circumstance that surrounded the Tsars. 'The Hermitage', housed in part of the Palace, is one of the richest art museums in the world. Having seen pictures of Tsar Nicholas with his sad resigned face sitting in the garden of Tsarskoe Selo on the outskirts of the city, just before he was sent out to Siberia with Tsarina Alexandra and children 'to be protected' and then shot so that no trace of the Romanovs would remain, we wanted to see the garden and asked the *Inturist* guide about it. He said he had not heard of that place. The name had been changed to Pushkin and that was why we had that firm negative.

The exquisite Pavlovsk Palace was still under repair — Leningrad and the surrounding area had suffered under German siege for 900 days from August 1941. Many died of starvation and cold. Much of the city was damaged or destroyed by air and artillery bombardment. At Pavlovsk, we saw artisans repairing the damage piece by piece. But the hundred or more French clocks of the eighteenth century were still keeping correct time. I remember a portrait in one of the rooms — a very beautiful aristocratic lady posing on a large polar-bear skin, and the same bear skin was laid out on the floor at the foot of the painting. She was Princess Yussoupov. Prince Yussoupov was one of those who assassinated Rasputin (December 1916). (Incidentally, the home of the Prince where the murder took place was pointed out to me in Leningrad and we passed by the Neva river where they broke the ice to conceal his body.) We did not see the Palace of Catherine the Great; it was under repair. The palace was known to be rather draughty and cold in the winter and Catherine was reputed to have called on her not inconsiderable palace guard to keep her warm!

We walked in the garden of Peter the Great's modest Summer Residence and were drenched by the hidden fountain which comes to life when stepped upon. Seeing all that he left behind, one feels that Peter the Great was magnificent not only in physical proportions but in his vision of a new Russia. The equestrian statue of Peter the Great — 'The Bronze Horseman' — is one of the important sights of this fascinating city. We did not miss the other sights — the Peter and Paul Fortress (1703-87) and its Cathedral where the Tsars were buried since the time of Peter the Great; the St. Isaac's Cathedral now turned into a museum of no significance except for its display of gold and silver decorations. We often walked along the Nevski Prospekt to admire the layout of the city. The many canals reminded one of Venice and the architecture was quite

Italian. The roadside stalls sold fruit which came from small private plots allowed to farmers. Indeed Leningrad still remains, in many respects, a Western city not only in appearance but also in the attitude and behaviour of its inhabitants. On the night of our arrival we had to go up to the roof restaurant of our hotel as it was too late to go into the dining room. The crowd we saw there could have come from a London, Paris, Madrid or Rome discotheque — noisy, unrestrained, calling for more drinks, getting up to dance in-between times — not the stolid appearance of the people in Moscow.

From Leningrad we came away with the feeling that the people living there still had a yearning for the past days when Leningrad, or rather Petrograd/St Petersburg, was the centre of Russia and one of the great centres of Europe.

This was too short and inadequate a visit for me to get any definite ideas or come to any definite conclusions. The international events following the Revolution had made it clear to Lenin and his followers that the first and foremost task for the USSR was to build its military strength to defend the Revolution in a deeply hostile world. They set out to do this by husbanding and concentrating its manpower and physical resources selectively. Today the Soviet Union is a super-power. Its achievements in aviation, nuclear physics and nuclear weapons, done on its own, are astonishing. For needs of lesser importance, it has turned to other countries — to Italy for *Fiat* cars for instance. For needs of advanced technology to exploit natural gas and petroleum, it waits for American or Japanese expertise and capital. This could well be said to be unbalanced. But looking at its past history it does make sense. In the process of development much deprivation, much hardship, much cruelty (*Gulag Archipelago*) has been caused. The concept of human rights has been given a different meaning. There can be no rights without obligations; these are two sides of the same coin. Western society emphasizes the rights, while the Communist society emphasizes the obligations. In the Communist philosophy, the State is supreme and the call of the people to be at the service of the State is a call which is superior to their own personal rights and conveniences. If they have suffered by a depressed standard of living compared to what exists in the Western countries, this is not to be regarded as a sacrifice but the discharge of their basic obligations to the State. (Despite Khruschev's boast, the per capita GNP per year according to World Bank calculations for 1975 gives the following picture: USSR — US$2,620; Federal Republic of Germany — $6,610; France — $5,760; Italy — $2,940; and Britain — $3,840.) The creation of a Consumer society is not the USSR's goal. The growing and disruptive industrial strife which is now a feature of Western society must have already confirmed them in their original philosophy.

These are two opposing systems of society. Which is better? If one finds comfort in the philosophy of Marcus Aurelius that 'A man needs little to lead a happy life' or accepts what Lord Bolingbroke once said, 'All our wants, beyond those which a very moderate income will supply, are purely imaginary', we who are in the other group may well have second thoughts. Perhaps, as in all things, the 'Middle Way' would be the ultimate answer. That would require working out, by a process of evolution, an equitable balance between rights and obligations, according to the circumstances in each country.

China Revisited

I HAVE DESCRIBED the general economic conditions in China after the war and just before the Communist Party under Mao Tse-tung took power.* I was keenly interested in observing what the Communist regime had accomplished since that time. The opportunity came in May 1977 when I accompanied the Director-General of FAO, Edouard Saouma, who was invited as an official guest by the Chinese Government. The transformation I saw was dramatic. In less than thirty years, the vast masses of the people, now officially estimated at 960 million, or almost a quarter of the world's population, have been provided with the basic necessities of life in food, clothing, housing, medical care, education and even old age security. How was this accomplished?

When the Communist Party took over in 1949, in its ideas and methods of approach to social and economic problems, it kept very close to the Soviet experience. This was reflected in the First Five-Year Plan (1953-57). The Plan gave the first priority to heavy industry and the second place to agriculture; agriculture was to serve industry. Planning was to be a central responsibility and material incentives were to be used to stimulate productivity. It soon became clear, however, that this Russian model of giving top priority to industry was not quite applicable to a country so predominantly agricultural as China. Mao Tse-tung demanded a re-examination of the whole approach in the light of the special conditions in China and the Party's own experience before it came into power. This demand found expression in the Great Leap Forward (1956-58) which saw intense self-criticism, launching of People's Communes, decentralisation of light and heavy industries (backyard furnaces, small fertilizer factories, hydro-electric generators) and emphasis on 'redness' rather than expertness. This brought about the break with the Soviet Union, with enormous consequences not only for China but for the whole world. The period that followed (1961-65) was one in which the wounds inflicted by some of the ill-conceived and ill-executed measures of the Great Leap Forward were allowed to heal and

*See *On the Road, China,* in Chapter 7.

the measures were withdrawn or rectified. This was also the period which saw the struggle between Mao Tse-tung and Liu Shou Chi, the latter advocating a return to the Russian model. Finally, as we all know, Liu Shou Chi lost and was discredited as a 'revisionist' and disappeared from the scene altogether. If one may indulge in some reflection here: it seems somewhat ironic that the term 'revisionist' should be applied to one who wanted to follow the conventional Russian line and not to one who changed the whole philosophy of communism as applied to China, though we know these words are common tools in communist disputation.

The China I saw was undoubtedly the creation of one person and that was Mao Tse-tung. Mao's belief, that the strength of a nation lay in the masses, and his determination to mobilize the masses in such a way that their strength might be a reality not just a mystique for politicians, provided the alchemy.

It was interesting to analyse how his ideas grew and were implemented. As in any other country, China had the old system of large and small landowners; the land was then the main capital. The approach to this problem was cautious. At the first stage (1949-54), land reform was undertaken to eliminate the old pattern of landlords and to give the land to the peasants. At the same time, Mutual Aid Teams were set up to take care of seasonal operations and projects for joint benefit. Private ownership of small farms was not disturbed. The next stage (1955-56) came when, still on the basis of private ownership, the land holdings were pooled for joint management. Income was distributed according to each participant's input of land, animals, tools and work. The third stage (1956-58) saw the individual ownership of land pass on to cooperatives, called Advanced Cooperatives. Income of individual members was no longer calculated by their capital inputs but by work points. It was not too long before the Advanced Cooperatives were found inadequate. The last stage was reached when Mao Tse-tung became convinced of the need to set up People's Communes merging on an average, ten advanced co-ops in each. Mao's philosophy of 'mass line' was the propelling force. The Great Proletarian Cultural Revolution (1966-73) was an expression of the same philosophy, though it spawned many political excesses and shook a whole country to its roots. The People's Commune concept showed its weakness as time passed because it was too far removed from work in the field. This led to the three-tier system:

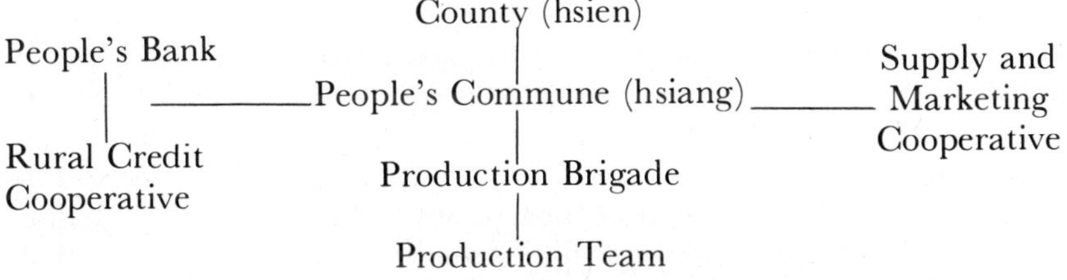

This three-tier system was evolved with the family household as the basic social unit. Production Teams were composed of 20 to 30 households and formed the basic production unit. It was also the unit from which the planning process started. The Production Brigade served largely as the planning and administrative unit. It received the reports and plans from the Production Teams before passing them on to the People's Commune. It was to undertake economic activities too large for a Team to handle, such as irrigation and land improvement works. It also handled agricultural processing plants, farm-machinery-repair shops, etc. It owned tractors and other farm machinery which was contracted out to the Production Teams. It operated schools and health stations. The People's Commune was to undertake activities too large for either of these, such as irrigation and drainage works, hydro-electric plants, farm machinery manufacture, etc. It was to run tractor stations, nurseries, experimental farms and was responsible for middle schools and hospitals.

This organizational structure was the end-result not of any rigid doctrinal approach but of years of trial and error, suited to the conditions in the countryside and was meant to maximise production. But what was equally important (perhaps more so) was the way that the decision-making process was decentralised and woven into this organization. The most important step was to decentralise planning, markedly deviating from the previous practice of central planning. At every hierarchical stage, opportunity was to be given to the people to make their voices heard. It was a two-way traffic, not a ready-made plan to be imposed from above. While there was a centrally directed national programme, much leeway was allowed to the lower units. The working of this method was facilitated by the presence of Party members at every level through the Revolutionary Committees and the 'correct line' was thus maintained.

In order that these intermediate bodies could deliberate policies and plans properly, each of them was to elect smaller streamlined bodies to represent their views. The People's Commune was to elect a hundred-member Council, the Brigades their own Revolutionary Committee and the Teams their leader and their leading group. The Chairman of the Commune's Revolutionary Committee (composed of about 20 members) was the Chief Executive of the Commune and allocated specific responsibilities to groups of members of the People's Council. I may say a word here about cooperatives that still remained side by side with the Commune (as shown in the figure). The Credit Cooperative implemented the State's monetary and credit policy under the direction of the People's Bank, mobilized rural savings and channelled the flow of agricultural and rural credit. The Supply and Marketing Cooperative distributed consumer goods obtained from the Wholesale Organization and the free-market products of the Commune members. (Each household was allowed private ownership of the house, some simple production tools, livestock such as a few pigs and a small plot of land.)

This is how the 'Mass Line' was put into its present form with as full a participation of the masses in the development process as possible.

The whole organizational structure was underpinned and reinforced in different ways. Decentralisation of industries, each tier being made responsible according to its capacity, was an important factor. The industries decentralised in this way related mainly to implements for farming and for processing farm products, thus avoiding the earlier mistake of the Great Leap Forward. This created extra employment, built up useful skills and utilised to the fullest extent the marketing possibilities of the local agricultural products. China is now pressing ahead with mechanisation but as yet it is in the preliminary stage.

At the same time, the rural masses were being assisted by more purposeful orientation of agricultural education, research and extension. All agricultural schools and colleges were directed to serve the needs of the peasants. Students and teachers were to spend, as a rule, three to four months in the year with farmers at the Commune level or to serve as productive labour in agro-industries. Veteran farmers were brought in as regular staff members to ensure that theory and practice were meaningfully related.

Research was decentralised and problem-oriented. Results of research were promptly included in production methods. There was to be no break between research and field operations. More reliance was to be placed on local insights to encourage coordination of research and extension.

Our visit covered practically all important aspects of these developments. We made a special trip to Tachai which Mao Tse-tung had held up as the agricultural model for all China *(Learn From Tachai)*.

With a population of only 470, of whom 160 were workers, the Tachai Brigade cultivated some 50 hectares, most of which the Brigade had terraced from the surrounding hills. The deep gullies which had been a major obstacle to cultivation had been filled up, the water flow being diverted through underground pipes and channels. This had been a major undertaking. The concrete pipes and tunnels testified to their skill and tenacity. All these works were done by the Brigade. Soil had to be hand-carried over long distances to increase productivity from 2 tons per hectare to 12 tons. Crops grown were barley, sorghum and millet.

What caught the imagination of all China and in particular of Chairman Mao was the spirit of the people of Tachai. A year or so after the various works had been completed there was a heavy rain flood which washed away all the terraces and concrete works so painstakingly done. It was a disaster of some magnitude. When the news spread, the Central Government at Peking offered help, as did the neighbouring countryside. But the people of Tachai stood firm. The teaching of Chairman Mao must be their guide. By *Hard Work and Self-reliance*, they must accomplish the task by themselves. This called for heroic effort —

physical as well as psychological. The work was undertaken again and successfully completed.

We were taken round by a local lady. Looking at her, one would think that she would be no different from any other young woman of her age. But she was steel all through. This 30-year old woman, Mrs. Kuou Fung-Lian, was Chairman of the Tachai Brigade and also Vice-Chairman of the County and Provincial Revolutionary Committees. She was one of the leaders in the Tachai reconstruction. She was decorated for her leadership as 'Hero of the Iron Brigade'. Here was a phenomenon. She symbolised the New China created by Mao Tse-tung. We also met Mr. Chen Yung-Kwei, one of the Vice-Premiers of China. He was one of the originators of the Tachai scheme. He began life as an illiterate peasant of this poverty-stricken area. Now, in spite of his many responsibilities in Peking, he made it a rule to spend a quarter of the year working in the fields with other members of the Brigade. One young member of our party, Vikram Shah, thus described his appearance: "a rugged face, heavily creased as if reflecting the terraces he had constructed on the land, and an irrestible smile which was not limited to his eyes and mouth but shone from his entire being".

All the householders lived in a barrack-type house with the kitchen in another similar house facing it. Both Chen Yung-Kwei and Kuou Fung-Lian lived in the same barracks with the others — one fairly spacious bed-sitting room and that was all. The beds were of bricks with a funnel underneath for heating — it would be sub-zero temperature in winter there. The furniture was sparse and bare. The only things to break the spartan austerity were the decorative photographs of Chairman Mao, Chairman Hua, Chou En-Lai and other dignitaries.

Among other memorable places we visited was the Chischashuan Prefecture, to see how they had managed to control the Haiho river which had always been a source of recurrent floods. In the past 20 years, 34 main canals, 4300 kilometres of dykes, 80 big and 1500 small reservoirs, and 490,000 pump wells had been completed. All these works had been carried out by nine Production Teams belonging to the local Brigade.

But very much more dramatic was the Red Flag Canal in the Linshien Prefecture. The trunk canal, 70 kilometres long, had been gouged out of solid rock. The ancillary works included 1500 kilometres of channels, 134 tunnels and 150 aqueducts. This gigantic work was done by 15 People's Communes with 480 Production Teams, supported by voluntary labour from other parts of the country. This was an example of the massive use of manpower in China. The Permanent Exhibition Hall in Peking enshrines this heroic effort in clay models and photographs. Some of the photographs show young men and women taking leave of their families for long absence as if going to war.

We did not see the political side of the New China nor any of the industrial complexes. We were kept strictly to our field of interest — agriculture and rural development. We did, however, get a glimpse of the political turmoil which was then going on, in the speeches of welcome and addresses in which the common theme was a condemnation of the 'Gang of Four'. In a primary school we were treated to a song by lovely red-cheeked five-year olds. The song was one of delight that Chairman Hua had smashed The Gang! To transform a nation, almost a quarter of the human race, within less than three decades, from a state of mass poverty frequented and decimated by famines into a society where there is no rich and poor but all have the basic needs of life, is surely an accomplishment which has no parallel in history. For this the world must pay homage to one man — Mao Tse-tung. Yet one hears the argument, specially from critics in affluent countries which pride themselves on free enterprise and the bounty that it brings, questioning whether this kind of mass mobilization behind closed doors has really been worthwhile. The answer must definitely be 'Yes'. For you must make sure that you live a decent life before you can think of higher things. No people are more gifted intellectually or artistically than the Chinese. This they have shown throughout their history — in art, architecture, philosophy and social behaviour. Such talent continues to flourish. Three decades is a very brief period in the history of a nation. This short period has laid a solid, sound and stable economic foundation for this great nation to build its future on. The present turmoil in China is no more than a natural reaction, for the creative urge of a people like the Chinese cannot be suppressed for too long. But did this change of direction require the levelling of the Great Helmsman to dust? Surely the leaders of China must realize that Mao Tse-tung's name is not written in water but in granite which will surely defy time!

Chapter 28

People I have met

Great Indian Leaders

IN WRITING MEMOIRS, the usual practice is to talk of personalities one has come across, and in doing so, to evaluate and to judge. I am not writing my memoirs, and even if I did my approach would be different. I would like to see whether and in what ways, consciously or unconsciously, people might have influenced me or left some permanent impression on me. The foundation of each individual's life is no doubt what he gets from heredity, but what one builds on that foundation comes largely from the environment in which one grows up — the early home life, the society in which one is brought up, and later in youth and maturity the problems one faces. In this process of growth, the people one comes across can play an important part.

In my boyhood and adolescence, the figure that stood out was my father. He was a doctor, teaching in a medical school in addition to his regular medical practice. He was a doctor of what might be called the 'old school'. He regarded the profession as one primarily for humanitarian service. He never asked for fees from his patients and took no interest in what he actually received. Once, when I was probably no more than ten years old he brought me with him to the post office. I carried a bag full of coins he had accumulated as fees over a period of time. When the coins were poured out before the Post Master, quite a number turned out to be counterfeit. This did not disturb my father, though it disturbed me. He did not show any sign of displeasure and apparently took it for granted. Sometimes he would be away from home for days which made my mother anxious. When he returned home it would come out that he had been on an emergency case some distance away. He was over 40 when he started using a bicycle. He fell off occasionally and was bruised and sore, yet his dogged persistence kept him going. He kept to an inexorable routine in his daily life, come hail, come storm — and it seemed to rain ceaselessly in that part of the world — arriving back home every time drenched to the skin. Once there was an epidemic of plague. The calls on his services were endless. He would come home after a punishing day, carefully leaving his shoes and over-clothes outside, and creep into the house. No wonder that our family seemed to have developed an immunity to plague and similar other not infrequent epidemics! At medical school he was greatly loved by his students — he was like a father to them all. I did not have a free and easy relationship with him because he was always busy. I do not remember him ever telling me how to behave or what to do. All that came from my mother. He once called me to say that I could get a scholarship, if I was willing to go to the famed Medical College in Calcutta. I said I preferred to wait and try for the prestigious Indian Civil Service (ICS) competitive examination held in London. He never referred to the subject again. When I was about to leave for Oxford, he handed me the entire amount that he thought I would need for my three-years' stay at University, provided I lived like a poor student. I respected his trust. I cannot calculate how much I owe to such a father.

When I returned from Oxford to take up my duties as a member of the ICS, I felt very humble, contrary to the general attitude of the members already working in the ICS and also the general opinion of my countrymen towards me. I did not look upon the ICS as 'rulers' but as servants of the people — to look after the needs and problems of the common people. I was aware of my inadequacy and therefore always willing to learn. I met different types of British officers in the Service — there were few Indians in the Service at that time — and only a few came up to the standards I had expected to find. I have briefly described one for whom I had great admiration — James Peddy, later shot by terrorists. Another of an entirely different kind was Sir William Prentice, Home Member in the Bengal Government: a formidable man, brusque, unapproachable, but one of the ablest in his day. I learnt that his wife was a patient in a mental home which I am sure affected him deeply and caused him to retreat into a granite-like shell. Once I went to the Club where he stayed to discuss a serious personal matter. He listened for a moment and then almost threw me out saying I should not have disturbed him on a Sunday. I was distressed, but realized from his attitude towards me later that he felt guilty and wanted to make amends. Beneath his stern exterior, he had a heart of gold, full of human kindness. He became a father-figure to me. His sudden death after emergency surgery was for me a deep personal sorrow.

I have mentioned Sir John Anderson's visit to Maldah where I was

District Officer. He was regarded in Britain as one of the most brilliant men of his time. The way he defended me against his Home Minister and the Police Chief for 'letting loose forces of anarchy and disorder' remains in my memory as an example of his integrity and strength as an administrator. After outstanding services to his country during the war as Lord President of the Council set up by Churchill, he once came to Rome on a private visit. I was asked by the British Ambassador to a luncheon to meet him. When luncheon was announced, he insisted that I should precede him as I was an Ambassador and he was only a private citizen. This was the correct protocol, but coming from one whom I had regarded as being very near to the gods, I felt very privileged.

I have mentioned my personal relationship with Lord Wavell who was Viceroy of India. I have sometimes wondered how it was that I was able to establish normal communication with such men as Prentice, Anderson and Wavell who were found so difficult and unapproachable by my British colleagues.

Two outstanding political figures with whom I came into close personal contact in my Bengal career were Dr. Shyamaprasad Mookherjee and Dr. B. C. Roy. I can never forget Shyamaprasad's speech in the Bengal Legislative Council in 1942 on the Bengal Famine. He described the scenes in the Calcutta streets and suburbs where people from the countryside were pouring in, desperate for food and dying in thousands. Dead bodies were piling up in the streets and the municipal services could not cope with this awful horror. He was a great orator, one of two, both Indians, whom I have not heard excelled. (The other was Srinivas Sastri, India's first High Commissioner to South Africa.) In 1921, he came to Oxford to address the students and the Master of Balliol presided. His style was a treat, reminiscent of Edmund Burke, with long passages beautifully balanced and perfectly finished. The Master of Balliol in his concluding address said that he had never known that English could be so eloquently spoken! Shyamaprasad was built on the lines of his father, big in size and big in every other way. Unfortunately, he chose his political base in the Hindu Mahasabha, a sectarian political party, and not the Congress which was nationwide. It was probably because of this that he never became an All-India figure like Jawaharlal Nehru. He was endowed with charismatic gifts and his early death left a void in Bengal and India that could never be filled.

Dr. B. C. Roy, one of the most eminent physicians in India at the time, also turned out to be a powerful political figure. I remember how back in 1919 in the first elections ever held in the country (under the Montagu-Chelmsford Reforms), he carried on his dramatic campaign to defeat S. N. Mullick, a protége of the British establishment. (To save embarrassment all round, the British Government immediately transferred Mullick to the India Council in London as a member — a very prestigious position in those days.) Dr. Roy as Chief Minister ruled the State like a dictator but respected all the panoply of democratic institutions.

He once asked me to attend one of his Cabinet meetings when I was on a visit to Calcutta from FAO. It was an interesting experience. The Ministers crept in quietly, seemingly in awe of the Master. Indeed, Dr. Roy treated them as his pupils in taking up and answering their questions and statements or giving them advice. At the same time, he did not stop anyone expressing himself. The question being discussed on that day was whether Bengal should merge itself with Bihar to form one linguistic State as recommended by the Commission set up by Nehru after the Madras riots over the introduction of Hinidi as the national language. In the course of the discussions, one of Dr. Roy's aides, his Judicial Secretary, a civil servent and not a Minister, went to the length of saying that he would resign if Dr. Roy gave his support to such a proposal — an extraordinary exhibition from a civil servant — but Dr. Roy did not show any resentment nor even any impatience! He was a kind, generous, big-hearted man. Even when holding his onerous position as Chief Minister of Bengal, he continued his old practice of setting apart one day in the week to freely see whoever came for medical advice or attention. This was in keeping with the story that once when he was sent to prison for anti-government activities, he became doctor to the jail staff as well as to the prisoners and the people in the neighbourhood. He had no thought for material gains for himself. But, unlike Shyamaprasad who was a rival in the political field, he was a master politician.

When I came to Delhi, I found an altogether different atmosphere. I have described the war tensions. I have also traced the conflicts of the political parties in the face of imminent transfer of power by the British. Since the beginning of the century, India had seen some great figures — intellectuals, thinkers, orators, statesmen and patriots all too ready to sacrifice their lives. I met some of that great band of men and women. The one I came to know best was Jawaharlal Nehru. My acquaintance with him grew deeper when I became Director-General of FAO at his personal initiative — the first non-Western to be Head of a UN Specialised Agency. Every year I came to New Delhi to see him and discuss the food and agriculture problems of the country. He would listen patiently and at the end would say "You see, I do not understand these problems, you should see Govinda Vallabh Panth." Sometimes he would relax and give me a glimpse of what was pressing on his mind. He was never dogmatic and always tried to see the other points of view. Of his daughter, Indira Ghandhi, he once said, "Indira wants to be a Minister, but I have sent her to the All-India Congress Committee and she is likely to be the next President." Perhaps even he did not realize that this experience in the All-India Congress would give her the insight into the Indian power politics which in future years would make her one of the most powerful Prime Ministers in Indian history. I met Nehru soon after the India-China War and I found him totally broken. His faith and his whole political philosophy lay shattered. Nehru often reminded me of what I had read about Pico della Mirandola, the legendary Renaissance scholar and philosopher of Florence who found elements of truth in all schools and thinkers and exalted the dignity of man. (No wonder that he was declared a heretic by Pope Innocent VIII for some of the discourses in his *Orations!*) Like Mirandola, Nehru was a true humanist. The essence of humanism, in the words of Walter Pater, the renowned nineteenth century art critic and essayist, is "the belief that nothing which has ever interested living men and women can wholly lose its vitality . . ."

Plato in his *Republic* defines the Philosopher-King as the ultimate 'ruler'. "Unless Philosopher-Kings become rulers, or rulers study philosophy, there will be no end of troubles for man". The only person I knew in modern times who answered Plato's ideal was Dr. Radhakrishnan who from a philosopher rose to be President of India. I came to know him when he was Professor of Indian Philosophy in Calcutta University in the 1940s. I had unbounded admiration for his great learning, and also as a man he was most charming. He wore his learning lightly. Everyone, young and old, had easy access to him. When I held charge of the troubled district of Midnapore in 1938, I invited him to lay the foundation stone of an edifice to perpetuate the memory of Tagore, one of the greatest sons of India who was born in that district. Aware of his torrential oratory, I lined up four shorthand experts to take down his speech. But it was to no purpose; his speed was too much for them and specially perplexing were his frequent quotations from the old Sanskrit texts. Many years later when I repeated this to him he told me a story, about when he was Professor of Oriental Philosophy at Oxford University. He was invited to preside over a school prize-giving ceremony. After he had spoken, he heard two elderly ladies on the front bench loudly whispering to each other "in what language did he speak?" I remained close to him since those early days. He came to Italy and Yugoslavia and then to Japan when I was Ambassador there. At Zagreb in Yugoslavia, two very young students came to see him. We were busy and I was hesitant, but I did not want to disappoint them. I took them in and left them with him for a while thinking that he would have a short session with them. When I came back, I found the boys in deep discussion with him and the Master showing equal enthusiasm. He could bring himself to discuss a question at any level of intelligence and with any age. He had the qualities of a great teacher. After he vacated his office as President of India, he fell ill and I went to see him. He raised himself on his bed to greet me. But after a few exchanges, I realized that some form of sclerosis was affecting his brain because his speech had become slurred and disjointed. Like the great composer Beethoven, bereft of hearing while some of his great works still lay unfinished, here lay the great philosopher bereft of his matchless gift of exposition!

It was my good fortune to have lived through the springtime of our nation. Like nature, a nation also seems to have its 'Four Seasons'. "If winter comes . . .?"

Leaders on the World Stage

MY DIPLOMATIC CAREER opened up further opportunities for me to meet some of the great personalities on the world stage. The initiative probably came from Nehru himself. I was then attending the UN General Assembly; I was asked to contact the Yugoslav representative at the General Assembly to ascertain if Yugoslavia would accept me as Ambassador if I was also concurrently accredited to Italy. The reply was immediate and positive, though the relations between Italy and Yugoslovia at that time were tense over the Trieste question. I was given a specially warm welcome when I called on Marshal Tito. This was soon after the break with the USSR. One reason I thought I received such a welcome was Tito's feeling of isolation and his desire to forge a link with India, then at the height of her international prestige with Nehru at the helm. Tito and his closest associates, Kardelj among others, gave freely of their time to explain and develop their thoughts, ideas and plans. I was aware that I roused the envy of the other Ambassadors in Belgrade. Sir Charles Peake, the British Ambassador, remarked that I should come more often from my base in Rome to Belgrade so that they could get more acquainted with Tito's thinking. Tito reciprocated my intense interest in the working of his young Communist State by facilitating my travels all over the country. I was able to see what I wanted. I came to realize how the leadership of one person could transform a whole nation. In spite of all that the people of the country had gone through, there was no sign of any despondency. People had tightened their belt as only Yugoslavs could and stood behind Tito as one man. Hope sustained their spirits with Tito there to guide the country. I have never known a parallel to this anywhere else.

I have mentioned Shigemitsu in Japan whose stoicism defying all misfortunes left a deep impression on my mind. De Gasperi was another whose leadership in war-shattered Italy was infused with a moral and spiritual quality rare in the political world of today. Harold Macmillan represented the élite of the British statesmen and hid his depth of understanding of human affairs with an easy manner. It was his *Winds of Change* speech during his African trip that ushered in independence to the British African colonies.

I first met John F. Kennedy in 1951 in Belgrade when he came as a member of a Congressional group to study the problems of Yugoslavia after its break with the USSR. He looked a rather sickly youth with freckles, a very junior member of the Congressional family. Who would have thought then that only a few years later he would be a world leader and open new frontiers! John Kennedy brought the gifts of youth — dreams and visions of a better world — to his high office. Could we ever forget the ring of his passionate sincerity when in his address inaugurating the 1963 World Food Congress he said, "We have the ability, as members of the human race, we have the means, we have the capacity, to eliminate hunger from the face of the earth in our lifetime. We need only the will."

Charles de Gaulle with his towering physique stood like a colossus astride the world. But he will always remain an enigma, a controversial figure even to his own countrymen. His biographers have failed to find the 'private face' of 'this man alone' who served his 'eternal France' but distrusted Frenchmen or rather the politicians representing Frenchmen. I have neither the competence nor the inclination to go into this controversy. But I know that when it came to purely human problems he was prepared to sit and listen because I was favoured in that way and in these circumstances he would show 'his private face'. When I paid my first call on him at the *Elysée,* I was forewarned by his aide that the President could spare only a few minutes for me. I was with him for nearly three-quarters of an hour to the discomfiture and unconcealed annoyance of his personal staff. For most of that time he was the listener. He used his interpreter only when he spoke. Before leaving, I asked him for a message on the occasion of the twentieth anniversary of the foundation of FAO. A few days later, an angry diplomat from the French Embassy in Rome came to my office and with a bang dumped the President's message on my table. This gesture of French displeasure was to show how wrong it was for me to have approached the President directly and not through the usual channels of the French Embassy (the Ambassador was a diplomatic colleague of mine in Tokyo!) and the *Quai d'Orsay.* (I had a similar experience in Washington when President Truman gave me an autographed photograph conveying his personal good wishes. I belonged to the profession of bureaucrats practically all my life, but it was only at that late period that I realized how we might appear in others' eyes.)

Since my boyhood years, I had known of Eamon de Valera, the great Irish revolutionary. He was about to be executed by the British for his part in the 1916 Uprising but was reprieved at the last moment. He escaped from Lincoln Gaol in 1919 and led his fellow revolutionaries in brilliant political campaigns as head of Sinn Féin and then of Fianna Fáil to lead the Irish people into a new era of freedom. His success as the President of the League of Nations Council and Assembly in 1931-32 is well known. In 1964 I was invited to Ireland to receive an Honorary degree at Dublin University — the University chartered by Queen Elizabeth in 1591 for 'the banishment of barbarism, tumults and disorderly living'! De Valera was then President of Ireland and I paid a long-desired call on him. In the earlier days, he had been the idol of the Indian revolutionaries. I mentioned to him the photograph I had seen in Los Angeles of Indian revolutionaries who had gathered in London to honour him. The great revolutionary was now the constitutional Head of the Irish Republic! He received me and my wife very kindly. He showed signs of approaching blindness. He leaves an indelible mark on history.

In one of my first visits to Egypt I called on President Nasser. He received me in the modest home he had lived in from his days as a Colonel in the Egyptian Army. Since his death he has come under attack from various quarters, even his close associate and successor Anwar Sadat saw the need to criticise him. The impression I had of him was of a very sincere upright man, a soldier with little political experience and therefore liable to sometimes take up politically impractical ideas. One example was the launching of the United Arab Republic by the federation of Egypt with Syria. Soon after this federation, I happened to visit Syria. At almost all levels — social, political, administrative — there was a feeling of bewilderment. The socialistic ideas adopted in Egypt did not apply to the prevailing conditions in Syria. Some of the leaders of the implacable Moslem Brotherhood were also in key positions of government at Damascus. In fact, very soon after, the federation broke up. (A similar attempt, in recent years, by Colonel Gadhafi, a disciple of Nasser, to bring about a union of Libya with Egypt failed to attract the interest of the more practical and realistic Anwar Sadat.) Nasser's approach to international problems, however, had one success. He joined with Nehru and Tito in the founding of the non-aligned movement which has now become a world force. Nasser gave me the impression of a man of undoubted integrity in the turbulent, tortuous world of Arab politics.

In Africa I came across two leaders of different types, but both impressive in their own ways. Nkrumah (Ghana) was a very attractive man with great personal charm. I postponed the opening of the first African Regional Conference deliberately until Ghana attained her independence so that I could have Nkrumah to preside. He received me in the Presidential Palace, once a colonial castle I believe. The castle suited his grand style. One remark he made jocously in his formal address was that all women in his country were his potential wives — a remark which the Ghanian ladies present seemed to enjoy enormously. It was indeed a great pity that one so charming and with so much promise should have met such a tragic end! I first came to know Julius Nyerere (Tanzania) when he stopped in Rome on his way back to his own country from abroad. His country was then on the verge of independence. I was lying ill at home. He came to see me. He impressed me deeply by his sincerity and generous idealism. One observation he made I remember in particular. He said that he would rather wait for some loose form of federation, economic and scientific, to take shape between the three East African countries — Tanzania, Kenya and Uganda — before indepdence. Nyerere has grown with the years. He was the first African leader I chose for the McDougall Lecture to the FAO Biennial Conference in 1963 — an important year for FAO as the mid-point of the Freedom from Hunger Campaign and the World Food Congress held in

Washington with its ringing Declaration.* Today he is a world leader. His address at the World Conference on Agrarian Reform and Rural Development in July 1979 was one of the most luminous expositions of the complex problems of rural development ever heard on a world platform. In his own country he is practising what he preaches. He is trying to establish a social order which gives equal opportunity to all. His military intervention in Uganda has come under some criticism but mostly by theoreticians. Anyone with knowledge of what Idi Amin had done to that country can only applaud.

There are moments in one's life when we feel too enmeshed in the cares of the world. My meeting with Swami Yogananda in California — alas, all too brief — gave me something I had not known before, not so much by what he said to me, because we had little time to ourselves, but by the atmosphere of his *ashram* and his godly presence. He knew that death was close but he did not spare himself in making my stay with him as full as possible. As the years pass, the memory of those brief moments becomes increasingly alive.

*See Appendix V on Page 316.

Appendices

Appendix I

Freedom from Hunger Campaign

Resolution No. 13/59, 27th October 1960

The Conference considering

(a) that a large part of the world's population still does not have enough to eat, and an even larger part does not get enough of the right kinds of food,

(b) that the increase in food production only barely exceeds population growth,

(c) that the increase in food production per capita is least marked in the less developed parts of the world,

(d) that food production in developed countries is being held back by limited marketing possibilities abroad and that even so, surpluses of some commodities have accumulated in some countries, and

(e) that under its Constitution FAO is the principal agency within the United Nations family of international agencies responsible for the encouragement of and aid to countries in raising levels of food production, consumption, and nutrition,

(1) Welcomes and approves the proposal for a Freedom from Hunger Campaign along the general lines suggested by the Director-General;

(2) Expresses appreciation of the co-operation in the Campaign promised by the United Nations and the specialised agencies;

(3) Authorises an international 'Freedom from Hunger Campaign' extending from 1960 through 1965, under the leadership and general co-ordination of FAO and with invitations to participate, as appropriate and approved by FAO, to (i) member countries of FAO; (ii) member countries of the United Nations and the United Nations specialised agencies, and the International Atomic Energy Agency (IAEA), and these agencies themselves; (iii) international non-governmental organizations that have established consultative relationships with FAO, the United Nations or the other

specialised agencies; (iv) religious groups; and (v) individuals and private organizations within the member countries specified in (i) and (ii) above;

(4) Emphasizes that the objectives of the Campaign can only be reached if the less developed countries develop effective and useful action projects to this end, and that the formulation and vigorous prosecution by them of such projects will increase the support for the Campaign in the more highly developed countries;

(5) Approves the creation of a Freedom from Hunger Campaign Trust Fund, to be administered in accordance with FAO's regulations, and for purposes and activities of the Campaign;

(6) Authorises the Director-General to appeal for voluntary contributions to:
(a) member countries as specified in para. 3 (i) and (ii);
(b) international non-governmental organizations;
(c) religious groups;
(d) private foundations or organizations in such member countries;

(7) (a) Authorises the Director-General, in the case of countries whose governments are not in a position to contribute directly to the Trust Fund, to discuss with these governments other ways in which they might be able to support the Campaign;
(b) Invites each member country to set up or utilise appropriate national bodies to promote and co-ordinate the Campaign in that country;
(c) Authorises the Director-General to carry on the Campaign with the funds available, in consultation with the Advisory Campaign Committee mentioned in para. 9 below;

(8) Authorises the Director-General to make preparations for a World Food Congress in 1963 immediately before the Twelfth FAO Conference Session, on the 20th anniversary of the Hot Springs Conference, when the Campaign will reach its climax;

(9) Establishes an Advisory Campaign Committee composed of the representatives of ten Member Nations to be designated by the Council, plus the chairmen of the Council, the Programme Committee and the Finance Committee, as *ex officio* members, this Committee to serve until the Eleventh Session of the Conference, with the following terms of reference: to advise and assist the Director-General in the development of a detailed programme for the Campaign, taking into account the suggestions made by the

Director-General to the Tenth Session of the Conference and the observations thereon by the Conference at that Session, and to report to the Council, and to establish a subcommittee of technical and economic experts on research needs and projects under the Campaign, selected for their competence and experience in various fields of work of FAO;

(10) Authorises the Director-General, after consultation with the Advisory Campaign Committee, to convene such meetings of representatives of governments or of such bodies mentioned in 7 (b) above as have been established, as may be considered by the Committee and the Director-General to be necessary or desirable, in order to review the progress and financial position of the Campaign;

(11) Authorises the Director-General to invite the non-governmental organizations specified in para. 3 (iii) above to participate in an advisory committee of non-governmental organizations, which shall on request consult with the Director-General and with representatives of other cooperating international organizations concerning plans for the Campaign and the activities of non-governmental organizations in assisting in the Campaign, at the same time providing an opportunity for the organizations represented to consult with one another;

(12) Requests the Director-General (a) to prepare reports to the Council concerning the detailed development of the Campaign and to present to the Conference in 1961 a detailed report on the current status of the Campaign and on proposed activities and their financial implications; and (b) to include in his financial proposals to the Eleventh Session of the Conference, separate provision for such funds as he may consider necessary to meet that portion of the FAO expenses for the Campaign for the 1962-63 biennium as cannot be covered by the Campaign Trust Fund;

(13) Requests the Council to keep the progress of the Campaign under review, to review reports from the Advisory Campaign Committee, and from the Programme and Finance Committees on the progress of the Campaign and its relation to the other work of FAO, and to present to the Conference its comments and suggestions on the further development of the Campaign.

Appendix II

Freedom from Hunger Campaign Basic Studies Series

The titles are as follows:

1. Weather and Food (WHO, Geneva, 1962)
2. Development through Food — a strategy for surplus utilisation (FAO, Rome, 1962)
3. Animal disease and human health (FAO, Rome, 1962)
4. Marketing — its role in increasing productivity (FAO, Rome, 1962)
5. Nutrition and working efficiency (FAO, Rome, 1962)
6. Education and training in nutrition (FAO, Rome, 1962)
7. Population and food supplies (United Nations, New York, 1962)
8. Aspects of economic development — the background to freedom from hunger (United Nations, New York, 1962)
9. Increasing food production through education, research and extension (FAO, Rome, 1963)
10. Possibilities of increasing world food production (FAO, Rome, 1963)
11. Third World Food Survey (FAO, Rome, 1963)
12. Malnutrition and disease (WHO, Geneva, 1963)
13. National development efforts (United Nations, New York, 1963)
14. Hunger and social policy (ILO, Geneva, 1963)
15. Education and agricultural development (UNESCO, Paris, 1963)
16. Wood: world trends and prospects (FAO, Rome, 1967)
17. Agriculture and industrialisation (FAO, Rome, 1967)
18. Agrarian reconstruction (FAO, Rome, 1968)
19. Fisheries in the food economy (FAO, Rome, 1968)
20. Forests, food and people (FAO, Rome, 1968)
21. Towards a strategy for agricultural development (FAO, Rome, 1969)
22. Nuclear techniques for increased food production (FAO/IAEA, Rome, 1969)
23. Agricultural development: a review of FAO's field activities (FAO, Rome, 1970)

Appendix III

United Nations General Assembly

Resolution 1714 (XVI), 19th December 1961

Provision of food surpluses to food-deficient peoples through the United Nations system, World Food Programme

THE GENERAL ASSEMBLY,

Recalling its Resolution 1496 (XV) of 27 October 1960 and Economic and Social Council Resolution 832 (XXXII) of 2 August 1960 on the provision of food surpluses to food-deficient peoples through the United Nations system,

Having considered the report of the Director-General of the Food and Agriculture Organization of the United Nations entitled *Development through Food – A strategy for surplus utilization,* the report of the Secretary-General on the role of the United Nations and the appropriate specialised agencies in facilitating the best possible use of food surpluses for the economic development of the less developed countries, and the joint proposal by the United Nations and the Food and Agriculture Organization of the United Nations regarding procedures and arrangements for multilateral utilisation of surplus food.

Having reviewed the action taken at the Eleventh Session of the Conference of the Food and Agriculture Organization on the utilisation of food surpluses and specifically, its Resolution stating that, subject to the concurrence of the General Assembly of the United Nations, an initial experimental programme for three years, to be known as the World Food Programme should be undertaken, having also noted in particular the reference to safeguards contained in paragraph 13 of the above mentioned Resolution of the Conference of the Food and Agriculture Organization,

Recognising the existing facilities for consultation provided by the Food and Agriculture Organization through its Consultative Subcommittee on Surplus Disposals,

Bearing in mind its Resolution 1710 (XVI) of 19 December 1961 on the United Nations Development Decade and, in particular, the reference in paragraph 4 (*d*) to the elimination of illiteracy, hunger and disease,

I

1. *Approves* the establishment of an experimental World Food Programme to be undertaken jointly by the United Nations and the Food and Agriculture Organization of the United Nations, in cooperation with other interested United Nations agencies and appropriate intergovernmental bodies, bearing in mind that the establishment of such a programme in no way prejudices the bilateral agreements between developed and developing countries, and accepts and endorses the purposes, principles, and procedures formulated in the first part of the Resolution approved by the Conference of the Food and Agriculture Organization on 24 November 1961, including the safeguards mentioned in that Resolution and General Assembly Resolution 1496 (XV) especially in paragraph 9 thereof;

2. *Approves* specifically the establishment of a United Nations/FAO Intergovernmental Committee of 20 States Members of the United Nations or members of the Food and Agriculture Organization to provide guidance on policy, administration and operations, and of a joint United Nations/FAO administrative unit reporting to the Secretary-General of the United Nations and the Director-General of the Food and Agriculture Organization;

3. *Requests* the Economic and Social Council at its resumed thirty-second session to elect, subject to the provisions of paragraph 9 below, ten States Members of the United Nations or members of the Food and Agriculture Organization to the United Nations/FAO Intergovernmental Committee, taking into account:

(*a*) the representation provided by the ten States that were elected to serve on the United Nations/FAO Intergovernmental Committee by the Council of the Food and Agriculture Organization,

(*b*) the need for balanced representation of economically developed and of developing countries, and other relevant factors such as the representation of potential participating countries, both contributing and recipient, equitable geographical distribution and the representation of both developed and less developed countries having commercial interests in international trade in foodstuffs, especially those highly dependent on such trade;

4. *Calls upon* the Economic and Social Council at its thirty-third session, in co-operation with the Council of the Food and Agriculture Organization, to review and to take appropriate action on the procedures and arrangements for the World Food Programme recommended by the United Nations/FAO Intergovernmental Committee;

5. *Decides* that pilot projects to be undertaken by the joint United Nations/FAO administrative unit, under the guidance of the United Nations/FAO Intergovernmental Committee, involving the use of food as an aid to economic and social development, shall be undertaken in

agreement between the Secretary-General, acting on behalf of the United Nations, and the Director-General, acting on behalf of the Food and Agriculture Organization;

6. *Concurs* in the calling of a conference where States Members of the United Nations or members of the Food and Agriculture Organization will be invited to pledge contributions;

7. *Requests* the Secretary-General, in co-operation with the Director-General of the Food and Agriculture Organization, to convene such a conference at United Nations Headquarters as soon as feasible after the concurrent sessions of the Economic and Social Council and the Council of the Food and Agriculture Organization;

8. *Urges* States Members of the United Nations or members of the Food and Agriculture Organization, in considering their contributions, to make every effort to assure the early attainment, on a voluntary basis, of the $100 million programme;

9. *Further requests* the Economic and Social Council, in co-operation with the Council of the Food and Agriculture Organization, at its next regular session following the pledging conference, to review the composition of the United Nations/FAO Intergovernmental Committee and to make by election for the balance of the three-year programme any adjustments of membership that might be deemed desirable in the light of the considerations outlined in paragraph 3 above;

10. *Instructs* the United Nations/FAO Intergovernmental Committee, in preparing recommendations on the conditions and procedures for the establishment and operation of the programme for the review and approval by the Economic and Social Council and the Council of the Food and Agriculture Organization, to proceed on the basis of the present resolution and the resolution of the Conference of the Food and Agriculture Organization of 24 November 1961, taking into account the joint proposal by the United Nations and the Food and Agriculture Organization of the United Nations regarding the procedures and arrangements for multilateral utilization of surplus food, statements made during the debates in the General Assembly and in the Conference of the Food and Agriculture Organization and such other conditions and procedures as may seem to it appropriate;

11. *Recommends* that Governments requesting assistance under this programme, the United Nations/FAO Intergovernmental Committee, and the joint United Nations/FAO administrative unit responsible for the administration of the programme keep the resident representatives fully informed about, and, within their field of competence, associated with, activities undertaken under the programme;

12. *Invites* the Secretary-General of the United Nations and the Director-General of the Food and Agriculture Organization to ensure that, in carrying out the programme, the joint United Nations/FAO administrative unit rely to the fullest extent possible on the existing staff and facilities of the United Nations and the Food and Agriculture

Organization, as well as other appropriate intergovernmental agencies;

13. *Requests* the United Nations/FAO Intergovernmental Committee to report annually to the Economic and Social Council and to the Council of the Food and Agriculture Organization on the progress made in the development of the programme, its administration and operation;

14. *Decides* to undertake, not later than at its nineteenth session, a general review of the programme, taking into account the objectives of its Resolution 1496 (XV);

II

Recognising that the experimental programme outlined above constitutes a step towards the broader objectives outlined in its Resolution 1496 (XV),

Recognising further that the ultimate solution to this problem of food deficiency lies in self-sustaining economic growth of the economies of the less developed countries to the point where they find it possible to meet their food requirements from their food-producing industries or from the proceeds of their expanding export trade,

Recognising that the effective utilisation of available surplus foodstuffs, in ways compatible with the principles of surplus disposal of the Food and Agriculture Organization of the United Nations, provides an important transitional means for relieving the hunger and malnutrition of food-deficient peoples, particularly in the less developed countries, and for assisting these countries in their economic development,

Recognising further that food aid is not a substitute for other types of assistance, in particular for capital goods,

1. *Recognises* that food aid to be provided under this programme should take into account other forms of assistance and country plans for economic and social development;

2. *Requests* the Secretary-General of the United Nations, in close co-operation with the Director-General of the Food and Agriculture Organization and with interested groups or agencies and jointly where appropriate, to undertake, as soon as is feasible, expert studies which would aid in the consideration of the future development of multilateral food programmes;

3. *Expresses the hope* that, in the light of these studies and of the experience gained, the progress of the experimental programme will be such as to permit the United Nations and the Food and Agriculture Organization to consider the possibility and advisability of increasing the programme, taking into account the advantages to developing countries, the interests of the contributing States, the interests of the food-exporting countries, the effectiveness of the programme and its contribution to the objectives of General Assembly Resolution 1496 (XV);

4. *Endorses again* the Freedom from Hunger Campaign launched by the Food and Agriculture Organization, and requests the Secretary-General of the United Nations and the Director-General of the Food and Agriculture Organization, simultaneously with the implementation of the present resolution, to pay particular attention to the necessity of improving and increasing local food production and to include, where appropriate, reference to this subject in the reports mentioned above, and requests the United Nations/FAO Intergovernmental Committee to consider the possibility of applying a reasonable proportion of resources resulting from the World Food Programme to this purpose.

1084th plenary meeting
19 December 1961

Appendix IV

Man's Right to Freedom from Hunger Manifesto

14 March 1963, Rome

More than half the human race is either undernourished or malnourished; yet about 150 billion dollars were spent on armaments in 1962, while the sum spent on development was an insignificant proportion of it. When we consider that in the twentieth century, one child out of three is born without any chance of living a normal life, we are forced to conclude that our civilisation is mutilating its human resources and reducing its chances of progress. The situation is getting worse because the population is increasing rapidly and food production is not keeping pace with it. The means are, nevertheless, at hand to meet this challenge and if they are used properly, the hope of a world free from the miseries of hunger can now be realized. Is mankind alive to this danger and prepared to meet it?

It is intolerable that the vast reservoir of knowledge and wealth which exists in the world is hardly being used for improving the lot of the many who are desperately in need of it. Of the several wants of man, food is primary. Hunger and malnutrition can impede the progress of a nation in every other sphere.

No development can be lasting which is not based on a mobilisation of national resources. But external aid is indispensable initially to guide and supplement these efforts. The impediments to improvement are social and economic rather than scientific and the supply of know-how and capital and the provision of facilities for education are the best means of ensuring an evolution towards a better life. The problems are complex, vast and urgent and can be solved only if national efforts are supported by international assistance and cooperation. In this connection, trade agreements should aim to preserve the dignity and independence of developing countries by enabling them to sell their products in the markets of the world. Cooperation by all economically advanced nations, both capitalist and communist, in the conquest of hunger and poverty, the common enemy of all mankind, may indeed breed sufficient mutual trust and confidence to assist progress towards that other of the fundamental freedoms, namely, freedom from the fear of war.

The Freedom from Hunger Campaign seeks to stimulate national and international effort. It aims to inform the Governments and educate the people so as to make the best use of the total resources of all nations.

We desire to state with all the emphasis at our command that freedom from hunger is man's first fundamental right. In order to achieve this, we suggest urgent and adequate national and international effort in which the Governments and the peoples are associated. More particularly, we desire to draw attention to the colossal waste of resources in the piling up of more and newer forms of armaments and the immense assistance to the campaign against hunger that even a partial diversion of these funds could achieve. We feel that international action for abolishing hunger will reduce tension and improve human relationships by bringing out the best instead of the worst in man.

Appendix V

Declaration of the World Food Congress

WE THE PARTICIPANTS of the World Food Congress assembled at Washington under the Freedom from Hunger Campaign to take the measure of the problems of hunger and malnutrition, and to explore the means for their solution,

having in mind that freedom from hunger is man's fundamental right and that all human beings — without distinction of any kind — are entitled to its realization through national effort and international cooperation,

conscious that today, in spite of twenty years of effort since the Hot Springs Conference which led to the foundation of FAO, the curse of hunger, malnutrition and poverty still afflicts more than half mankind,

alarmed by the extent to which the explosive growth of population, at a rate unmatched by adequate increases in productivity, is intensifying human needs and is giving still greater urgency to the attainment of freedom from hunger,

profoundly aware that the recent attainment of political independence by many hundred millions of the world's population gives a new urgency and new dimension to the aspiration for higher levels of living, of which freedom from hunger is the first prerequisite,

convinced that scientific and technological progress now make it possible to free the world from hunger, but that such freedom can only be accomplished if all the available human and natural resources of the world are mobilised to this end through balanced economic and social development,

HEREBY DECLARE

that the persistence of hunger and malnutrition is unacceptable morally and socially, is incompatible with the dignity of human beings and the equality of opportunity to which they are entitled, and is a threat to social and international peace;

that the elimination of hunger is a primary task of all men and women, who must recognise their duties as well as their rights as members of the human race, and must fight to achieve freedom from hunger in every corner of the earth; this obligation being also inherent in the pledge of the nations under the United Nations Charter to take joint and separate action, to achieve higher standards of living, full employment and conditions of economic and social progress and development as indispensable elements of peace;

that the responsibility to free the world from the scourge of hunger lies jointly

with the developing countries themselves who must take all measures within their power which are necessary to achieve this objective;

with the developed nations who must co-operate with the developing countries in their efforts, realizing that freedom from hunger cannot long be secure in any part of this interdependent world unless it is secure in all the world;

with the United Nations and the specialised agencies who must intensify and coordinate their efforts to assist the nations in this task;

with other international organizations and with non-governmental organizations, e.g. religious, youth, women's organizations and other voluntary groups, agricultural and labour organizations and associations of trade and industry, who must inform and stimulate the people so that they can play their part with understanding and vigour;

THEREFORE URGE

that the task of elimination of hunger from the face of the earth should be conceived in the framework of a worldwide development dedicated to the fullest and most effective use of all human and natural resources, to ensure a faster rate of economic and social growth, and

that to this effect, speedy and decisive action be taken:

(1) by all governments of the developing countries

(a) for a planned and integrated use of resources which at present are largely underutilised;

(b) for the adaptation of their institutions to the requirements of economic and social progress; and, more specifically, to secure the most effective administrative machinery, to give incentives to increased production through ensuring just and stable prices, and to reform, where required, unjust and obsolete structures and systems of land tenure and land use so that the land might become, for the man who works it, the basis of his economic betterment, the foundation of his increasing welfare, and the guarantee of his freedom and dignity;

(2) for the maximum utilisation of the stock of scientific and technical knowledge and the promotion of both short- and long-term adaptive research suited to the conditions and requirements of the developing countries;

(3) for the massive and purposive education of the rural populations, so that they will be capable of applying modern techniques and systems, and for universal education to expand the opportunities for all:

FURTHER URGE

that to assist national efforts, and allow speedier implementation of development programmes within a worldwide framework, international cooperation be strengthened, in particular so that

(1) present adverse and disturbing tendencies in the trade of the developing countries be reversed and that for that purpose adequate and comprehensive commodity agreements be devised, development plans be coordinated and other appropriate measures taken, and

(2) the volume and effectiveness of financial, material and technical assistance be increased, and

(3) there be a more equitable and rational sharing of world abundance, including an expanded and improved utilisation of food surpluses for the purpose of economic and social development:

EXPRESS THE HOPE

that the current efforts for bringing about universal disarmament will succeed and that the vast sums now being spent on instruments of destruction will become increasingly available for the elimination of hunger and malnutrition and the promotion of human well-being:

THEREFORE PLEDGE OURSELVES

and highly resolve to take up the challenge of eliminating hunger and malnutrition as a primary task of this generation, thus creating basic conditions for peace and progress for all mankind;

to mobilise every resource at our command to awaken world opinion and to stimulate all appropriate action, public and private, national and international, for this overriding task, and to this end give our wholehearted support to the Freedom from Hunger Campaign until its final goal is achieved.

Resolution of the World Food Congress

THE WORLD FOOD CONGRESS, considering that the success of the Freedom from Hunger Campaign depends on the active cooperation of the government and the people of every country, recommends:

(1) the placing of all national Freedom from Hunger Campaign committees on a continuing basis;

(2) the creation of effective national committees in all states and territories members of FAO and of other organizations of the United Nations system; and

(3) holding of a World Food Congress periodically to review a world survey, presented by the Director-General of FAO, of the world food situation in relation to population and overall development, together with a proposed programme for future action.

Appendix VI

Young World Manifesto

16 October 1965, Rome

TO ALL YOUNG PEOPLE, everywhere, from the Young World Assembly meeting in Rome.

Half the world does not have enough to eat. Each year, as a result, many millions die young, as surely as if shot by the guns of a tyrant. Many more are maimed for life by hunger, in body or in spirit.

We say to you, this suffering can and must be stopped. When all of us, whether we live with it or far away in the rich, well-fed countries, make up our minds to end this hunger, we can do it.

The earth is ruled mainly by people out of touch with the young world. They know that men starve and die in millions. But they think it more important to make guns, bombs, warships, rockets, to send us to fight one another, than to provide seed and water, schools and hospitals, so that we might feed and serve one another.

Twenty years ago today, men of foresight set up the Food and Agriculture Organization of the United Nations, to lead the attack on hunger. Many eat better than they would have done without it. Yet, after 20 years, there are more hungry people than ever before. In another 20 years' time, if we do not act, there will be yet more, famine will haunt many lands and we shall be fighting one another again. We must prevent such an outcome through the mobilisation of the young world.

Know your power and know what you must do.

If you live in a rich country, you have wealth to share. Tell your fellow countrymen about the hunger in other lands. Demand of your governments that much more of the nations' wealth — very many billions of dollars' worth — should go to world development.

If you live in a poor country, demand adequate food for your fellows. Do not turn your backs on the land and people who provide the food; instead, work with them for rural development. Plan with them, so that starting with what little they have, they themselves can develop in body and in spirit.

If you are educated in special knowledge and skills, do not accept the old priorities. Know that science and technology, that can send men into space, need only to be released into the poor lands to work even greater miracles. See that your skills are used to help the needy.

If you are a young parent, resolve to end the suffering of children. Know, too, how to plan the size of your family, so that the progress of all is not compromised. Let us all make plain to the rulers that the division of the world into rich and poor must end and that we know that efforts equivalent to the many billions of dollars wasted on armaments are needed to develop the world. Let them know, too, that if political or financial systems prevent a just distribution of food and wealth, those systems must be replaced.

Above all we must show our willingness to work for world development, and demand that we be given the opportunity to do so. Mankind is one family in which each of us has the duty to help the other.

We who are meeting at the Young World Assembly have pledged ourselves to this struggle as countless other young people all over the world have done. Our generation has power and knowledge that no previous generation has ever had. With these we must create a world in which the human spirit is set free from hunger and want, for ever.

Appendix VII

General Assembly — Sixth Special Session
Resolution adopted on the Report of the *Ad Hoc* Committee

Declaration on the Establishment of a New International Economic Order

Resolution No. 3201 (S-VI), Adopted on 1 May 1974

WE, THE MEMBERS OF THE UNITED NATIONS,

Having convened a special session of the General Assembly to study for the first time the problems of raw materials and development, devoted to the consideration of the most important economic problems facing the world community,

Bearing in mind the spirit, purposes and principles of the Charter of the United Nations to promote the economic advancement and social progress of all peoples,

Solemnly proclaim our united determination to work urgently for THE ESTABLISHMENT OF A NEW INTERNATIONAL ECONOMIC ORDER based on equity, sovereign equality, interdependence, common interest and co-operation among all States, irrespective of their economic and social systems which shall correct inequalities and redress existing injustices, make it possible to eliminate the widening gap between the developed and the developing countries and ensure steadily accelerating economic and social development and peace and justice for present and future generations, and, to that end, declare:

(1) The greatest and most significant achievement during the last decades has been the independence from colonial and alien domination of a large number of peoples and nations which has enabled them to become members of the community of free peoples. Technological progress has also been made in all spheres of economic activities in the last three decades, thus providing a solid potential for improving the well-being of all peoples. However, the remaining vestiges of alien and colonial domination, foreign occupation, racial discrimination, apartheid and neo-colonialism in all its forms continue to be among the greatest obstacles to the full emancipation and progress of the developing countries and all the peoples involved. The benefits of technological progress are not shared equitably by all members of the international community. The developing countries, which constitute 70 per cent of the world's population, account for only 30 per cent of the world's income. It has proved impossible to achieve an even and balanced development of the international

community under the existing international economic order. The gap between the developed and the developing countries continues to widen in a system which was established at a time when most of the developing countries did not even exist as independent States and which perpetuates inequality.

(2) The present international economic order is in direct conflict with current developments in international political and economic relations. Since 1970, the world economy has experienced a series of grave crises which have had severe repercussions, especially on the developing countries because of their generally greater vulnerability to external economic impulses. The developing world has become a powerful factor that makes its influence felt in all fields of international activity. These irreversible changes in the relationship of forces in the world necessitate the active, full and equal participation of the developing countries in the formulation and application of all decisions that concern the international community.

(3) All these changes have thrust into prominence the reality of interdependence of all the members of the world community. Current events have brought into sharp focus the realization that the interests of the developed countries and those of the developing countries can no longer be isolated from each other, that there is a close interrelationship between the prosperity of the developed countries and the growth and development of the developing countries, and that the prosperity of the international community as a whole depends upon the prosperity of its constituent parts. International cooperation for development is the shared goal and common duty of all countries. Thus the political, economic and social well-being of present and future generations depends more than ever on cooperation between all the members of the international community on the basis of sovereign equality and the removal of the disequilibrium that exists between them.

(4) The new international economic order should be founded on full respect for the following principles:

(a) Sovereign equality of States, self-determination of all peoples, inadmissibility of the acquisition of territories by force, territorial integrity and non-interference in the internal affairs of other States;

(b) The broadest cooperation of all the States Members of the international community, based on equity, whereby the prevailing disparities in the world may be banished and prosperity secured for all;

(c) Full and effective participation on the basis of equality of all countries in the solving of world economic problems in the common interest of all countries, bearing in mind the necessity to ensure the accelerated development of all the developing countries, while devoting particular attention to the adoption of special measures in favour of the least developed, land-locked and island developing countries as well as those developing countries most seriously affected by economic crises and natural calamities, without losing sight of the interests of other developing countries;

(d) The right of every country to adopt the economic and social system that it deems the most appropriate for its own development and not to be subjected to discrimination of any kind as a result;

(e) Full permanent sovereignty of every State over its natural resources and all economic activities. In order to safeguard these resources, each State is entitled to exercise effective control over them and their exploitation with means suitable to its own situation, including the right to nationalisation or transfer of ownership to its nationals, this right being an expression of the full permanent sovereignty of the State. No State may be subjected to economic, political or any other type of coercion to prevent the free and full exercise of this inalienable right;

(f) The right of all States, territories and peoples under foreign occupation, alien and colonial domination or apartheid to restitution and full compensation for the exploitation and depletion of, and damages to, the natural resources and all other resources of these States, territories and peoples;

(g) Regulation and supervision of the activities of transnational corporations by taking measures in the interest of the national economies of the countries where such transnational corporations operate on the basis of the full sovereignty of those countries;

(h) The right of the developing countries and the peoples of territories under colonial and racial domination and foreign occupation to achieve their liberation and to regain effective control over their natural resources and economic activities;

(i) The extending of assistance to developing countries, peoples and territories which are under colonial and alien domination, foreign occupation, racial discrimination or apartheid or are subjected to economic, political or any other type of coercive measures to obtain from them the subordination of the exercise of their sovereign rights and to secure from them advantages of any kind, and to neo-colonialism in all its forms, and which have established or are endeavouring to establish effective control over their natural resources and economic activities that have been or are still under foreign control;

(j) Just and equitable relationship between the prices of raw materials, primary commodities, manufactured and semi-manufactured goods exported by developing countries and the prices of raw materials, primary commodities, manufactures, capital goods and equipment imported by them with the aim of bringing about sustained improvement in their unsatisfactory terms of trade and the expansion of the world economy;

(k) Extension of active assistance to developing countries by the whole international community, free of any political or military conditions;

(l) Ensuring that one of the main aims of the reformed international monetary system shall be the promotion of the development of the developing countries and the adequate flow of real resources to them;

(m) Improving the competitiveness of natural materials facing competition from synthetic subsitutes;

(n) Preferential and non-reciprocal treatment for developing countries, wherever feasible, in all fields of international economic cooperation whenever possible;

(o) Securing favourable conditions for the transfer of financial resources to developing countries;

(p) Giving to the developing countries access to the achievements of modern science and technology, and promoting the transfer of technology and the creation of indigenous technology for the benefit of the developing countries in forms and in accordance with procedures which are suited to their economies;

(q) The need for all States to put an end to the waste of natural resources, including food products;

(r) The need for developing countries to concentrate all their resources for the cause of development;

(s) The strengthening, through individual and collective actions, of mutual economic, trade, financial and technical cooperation among the developing countries, mainly on a preferential basis;

(t) Facilitating the role which producers' associations may play within the framework of international cooperation and, in pursuance of their aims, *inter alia* assisting in the promotion of sustained growth of the world economy and accelerating the development of developing countries.

(5) The unanimous adoption of the International Development Strategy for the Second United Nations Development Decade was an important step in the promotion of international economic cooperation on a just and equitable basis. The accelerated implementation of obligations and commitments assumed by the international community within the framework of the Strategy, particularly those concerning imperative development needs of developing countries, would contribute significantly to the fulfilment of the aims and objectives of the present Declaration.

(6) The United Nations as a universal organization should be capable of dealing with problems of international economic cooperation in a comprehensive manner and ensuring equally the interests of all

countries. It must have an even greater role in the establishment of a new international economic order. The Charter of Economic Rights and Duties of States, for the preparation of which the present Declaration will provide an additional source of aspiration, will constitute a significant contribution in this respect. All the States Members of the United Nations are therefore called upon to exert maximum efforts with a view to securing the implementation of the present Declaration, which is one of the principal guarantees for the creation of better conditions for all peoples to reach a life worthy of human dignity.

(7) The present Declaration on the Establishment of a New International Economic Order shall be one of the most important bases of economic relations between all peoples and all nations..

2229th Plenary Meeting
1 May 1974

Appendix VIII

Declaration of Principles

The World Conference on Agrarian Reform and Rural Development

(1) Having met from 12 to 20 July 1979 in Rome, Italy,

(2) Recalling that the Conference was a continuation of a long and deep concern of the international community with agrarian and rural questions,

(3) Recalling also previous United Nations conferences, particularly the World Food Conference of 1974, the VI and VII Special Sessions of the UN General Assembly and the Charter of Economic Rights and Duties of States,

(4) Recognising that most development efforts have not yet succeeded in satisfying the aspirations of peoples and their basic requirements consistent with principles of human dignity and international social justice and solidarity, especially in the rural areas of developing countries,

(5) Conscious that past development efforts and programmes have largely failed to reach and adequately benefit the rural areas and have in many cases contributed to urban-rural imbalance in development, neglected the dynamism and diversity of authentic cultural values of the rural population and led to imbalances within the rural sector,

(6) Aware of the need to adopt appropriate population policies within the context of socio-economic development, achieve ecological harmony and conserve finite resources,

(7) Believing that poverty, hunger and malnutrition retard national development efforts and negatively affect world social and economic stability and that their eradication is the primary objective of world development,

(8) Convinced that agrarian reform is a critical component of rural development and that the sustained improvement of rural areas, in the context of promotion of national self-reliance and the building of the New International Economic Order, requires fuller and more equitable access to land, water and other natural resources; widespread sharing of economic and political power; increasing and more productive employment; fuller use of human skills and

energies; participation and integration of rural people into the production and distribution systems; increased production, productivity and food security for all groups; and mobilisation of internal resources,

(9) Reaffirming the UN General Assembly resolutions on world peace and disarmament and resolutions 3201 and 3202 of the VI Special Session relating to efforts "to put an end to all forms of foreign occupation, racial discrimination, apartheid, colonial, neo-colonial and alien domination and exploitation through the exercise of permanent sovereignty over all natural resources" and recognising their bearing on agrarian reform and rural development,

(10) Aware that, while the primary responsibility for agrarian reform and rural development rests with individual governments, a sustained and global programme will demand strong political commitment, active cooperation within the international community and commitment and effective use of financial, technical and human resources in a sound, systematic and coordinated manner,

(11) Affirming that such cooperation must be based on resolute adherence to principles of independence, national sovereignty, self-determination of peoples and non-intervention in the internal affairs of States,

(12) Recalling the relevant resolution of the UN General Assembly at its 33rd Session that the new International Development Strategy should provide a set of interrelated and concerted measures in all sectors of development in order to promote the economic and social development of the developing countries and to ensure their equitable, full and effective participation in the formulation and application of all decisions in the field of development and international economic cooperation,

(13) Emphasizing that measures to strengthen international cooperation in agrarian reform and rural development are most effective when national strategies fully recognise the interdependence of industry and agriculture,

(14) Recognising that the UN system has a responsibility to formulate a new International Development Strategy and that the Food and Agriculture Organization, under the terms of its constitution, has an explicit obligation to elaborate those parts of this new strategy in regard to food, agriculture, nutrition and other areas of its competence, and should play a leading role in assisting developing countries to promote agrarian reform and rural development,

(15) *HEREBY DECLARES* that a Programme of Action should be founded on the following guidelines and principles:

(a) that the fundamental purpose of development is individual and social betterment, development of endogenous capabilities and improvement of the living standards of all people, in particular the rural poor;

(b) that the right of every State to exercise full and permanent sovereignty over its natural resources and economic activities and to adopt the necessary measures for the planning and management of its resources is of vital importance to rural development;

(c) that the use of foreign investments for agricultural development of developing countries, in particular that of transnational corporations, must be in accordance with national needs and priorities;

(d) that national progress based on growth with equity and participation requires a redistribution of economic and political power, fuller integration of rural areas into national development efforts, with expanded opportunities of employment and income for rural people, and development of farmers' associations, cooperatives and other forms of voluntary autonomous democratic organizations of primary producers and rural workers;

(e) that appropriate population policies and programmes can contribute to long-term social and economic progress;

(f) that maximum efforts should be made to mobilise and use productively domestic resources for rural development;

(g) that governments should introduce positive bias in favour of rural development and provide incentives for increased investment and production in rural areas;

(h) that equitable distribution and efficient use of land, water and other productive resources, with due regard for ecological balance and environmental protection, are indispensable for rural development, for the mobilisation of human resources and for increased production for the alleviation of poverty;

(i) that diversification of rural economic activities, including integrated crop-livestock development, fisheries and aquaculture and integrated forestry development, is essential for broad-based rural development;

(j) that location of industries in the rural areas, in both the public and private sectors and particularly agro-industries, provides necessary and mutually reinforcing links between agriculture and industrial development;

(k) that policies and programmes affecting agrarian and rural systems should be formulated and implemented with the full understanding and participation of all rural people, including youth, and of their own organizations at all levels, and that development efforts should be responsive to the varying needs of different groups of rural poor;

(l) that understanding and awareness of the problems and opportunities of rural development among people at all levels and that improving the interaction between development personnel and the masses through an efficient communication system are prerequisites for the success of rural development strategy;

(m) that constant vigilance should be kept to ensure that benefits of agrarian reform and rural development are not offset by the reassertion of past patterns of concentration of resources in private hands or by the emergence of new forms of inequity;

(n) that women should participate and contribute on an equal basis with men in the social, economic and political processes of rural development and share fully in improved conditions of life in rural areas;

(o) that international cooperation should be strengthened and a new sense of urgency introduced to augment the flow of financial and technical resources for rural development;

(p) that all governments should undertake new and more intensive efforts to ensure world food security and overcome inequities and instability in the trade of agricultural commodities of particular importance to developing countries; and

(q) that developing countries, with the support of international development organizations, should strengthen their technical cooperation in rural development and foster policies of collective self-reliance.

Appendix IX

Vote of Thanks to Dr. B. R. Sen

FAO Conference Resolution No. 33/67
Adopted 22 February 1967

THE CONFERENCE *recognising* the eminent services rendered by Dr. B. R. Sen to the Organization in raising it to the standing it now enjoys by his exceptional energy, devotion to duty and great competence in development questions,

Considering the unique contribution he has made in creating among the leaders and peoples of the world an understanding of the problems of hunger and malnutrition and of the imperative need to accelerate agricultural and general economic development,

Noting that under Dr. Sen's leadership FAO has taken major steps toward solving those problems, as exemplified by the Freedom from Hunger Campaign, the World Food Programme and work on the Indicative World Plan,

Conveys to him its warmest thanks and congratulations for his magnificent achievements that will be lastingly remembered; and

Hopes that he may continue to serve the ideals for which FAO stands throughout the world.

The Conference *decides* to have this resolution inscribed on parchment, signed by the Chairman of the Conference and presented to Dr. Sen.

The Conference *decides further* to create a new permanent feature of FAO activity associated with the name of Dr. Sen and directed towards the goals he served so well; and

Requests the Council to take action as appropriate to constitute such a permanent and continuing tribute to Dr. Sen.

Index

Index